The IMA Volumes in Mathematics and Its Applications

Volume 6

Series Editors
George R. Sell   Hans Weinberger

T0325107

Institute for Mathematics and Its Applications
IMA

The **Institute for Mathematics and Its Applications** was established by a grant from the National Science Foundation to the University of Minnesota in 1982. The IMA seeks to encourage the development and study of fresh mathematical concepts and questions of concern to the other sciences by bringing together mathematicians and scientists from diverse fields in an atmosphere that will stimulate discussion and collaboration.

The IMA Volumes are intended to involve the broader scientific community in this process.

Hans Weinberger, Director
George R. Sell, Associate Director

IMA Programs

1982–1983 Statistical and Continuum Approaches to Phase Transition

1983–1984 Mathematical Models for the Economics of Decentralized Resource Allocation

1984–1985 Continuum Physics and Partial Differential Equations

1985–1986 Stochastic Differential Equations and Their Applications

1986–1987 Scientific Computation

1987–1988 Applied Combinatorics

1988–1989 Nonlinear Waves

Springer Lecture Notes from the IMA

The Mathematics and Physics of Disordered Media
Editors: Barry Hughes and Barry Ninham
(Lecture Notes in Mathematics, Volume 1035, 1983)

Orienting Polymers
Editor: J. L. Ericksen
(Lecture Notes in Mathematics, Volume 1063, 1984)

New Perspectives in Thermodynamics
Editor: James Serrin
(Springer-Verlag, 1986)

Models of Economic Dynamics
Editor: Hugo Sonnenschein
(Lecture Notes in Economics, Volume 264, 1986)

Constantine Dafermos, J.L. Ericksen,
and David Kinderlehrer
Editors

Amorphous Polymers and Non-Newtonian Fluids

With 30 Illustrations

Springer-Verlag
New York Berlin Heidelberg
London Paris Tokyo

Constantine Dafermos
J.L. Ericksen
David Kinderlehrer
The Institute for Mathematics and Its Applications
Minneapolis, Minnesota 55455

AMS Classification: 76 A 05

Library of Congress Cataloging in Publication Data
Amorphous polymers and non-Newtonian fluids.
(The IMA volumes in mathematics and its applications ; v. 6)
Proceedings of one of a series of IMA workshops held during 1984–85.
Bibliography: p.
Includes index.
1. Non-Newtonian fluids—Congresses. 2. Polymers and polymerization—Congresses. I. Dafermos, C. M. (Constantine M.) II. Ericksen, J. L. (Jerald L.), 1924– III. Kinderlehrer, David. IV. University of Minnesota. Institute for Mathematics and Its Applications. V. Series.
QA929.5.A47 1987 532 87-12990

©1987 by Springer-Verlag New York Inc.
All rights reserved. This work may not be translated or copied in whole or in part without the written permission of the publisher (Springer-Verlag, 175 Fifth Avenue, New York, New York 10010, U.S.A.), except for brief excerpts in connection with reviews or scholarly analysis. Use in connection with any form of information storage and retrieval, electronic adaptation, computer software, or by similar or dissimilar methodology now known or hereafter developed is forbidden.
The use of general description names, trade names, trademarks, etc. in this publication, even if the former are not especially identified, is not to be taken as a sign that such names, as understood by the Trade Marks and Merchandise Marks Act, may accordingly be used freely by anyone.
Permission to photocopy for internal or personal use, or the internal or personal use of specific clients, is granted by Springer-Verlag New York Inc. for libraries registered with the Copyright Clearance Center (CCC), provided that the base fee of $0.00 per copy, plus $0.20 per page is paid directly to CCC, 21 Congress Street, Salem, MA 01970, U.S.A. Special requests should be addressed directly to Springer-Verlag New York, 175 Fifth Avenue, New York, New York 10010, U.S.A.
96556-4/1987 $0.00 + .20

Printed and bound by Edwards Brothers, Ann Arbor, Michigan.
Printed in the United States of America.

9 8 7 6 5 4 3 2 1

ISBN 0-387-96556-4 Springer-Verlag New York Berlin Heidelberg
ISBN 3-540-96556-4 Springer-Verlag Berlin Heidelberg New York

The IMA Volumes in Mathematics and Its Applications

Current Volumes:

Volume 1: Homogenization and Effective Moduli of Materials and Media
Editors: Jerry Ericksen, David Kinderlehrer, Robert Kohn, and J.-L. Lions
Volume 2: Oscillation Theory, Computation, and Methods of Compensated Compactness
Editors: Constantine Dafermos, Jerry Ericksen, David Kinderlehrer, and Marshall Slemrod
Volume 3: Metastability and Incompletely Posed Problems
Editors: Stuart Antman, Jerry Ericksen, David Kinderlehrer, and Ingo Müller
Volume 4: Dynamical Problems in Continuum Physics
Editors: Jerry Bona, Constantine Dafermos, Jerry Ericksen, and David Kinderlehrer
Volume 5: Theory and Applications of Liquid Crystals
Editors: Jerry Ericksen and David Kinderlehrer
Volume 6: Amorphous Polymers and Non-Newtonian Fluids
Editors: Constantine Dafermos, Jerry Ericksen, and David Kinderlehrer
Volume 7: Random Media
Editor: George Papanicolaou
Volume 8: Percolation Theory and Ergodic Theory of Infinite Particle Systems
Editor: Harry Kesten

Forthcoming Volumes:

1985–1986: Stochastic Differential Equations and Their Applications
Hydrodynamic Behavior and Interacting Particle Systems
Stochastic Differential Systems, Stochastic Control Theory and Applications

1986–1987: Scientific Computation
Computational Fluid Dynamics and Reacting Gas Flows
Numerical Algorithms for Modern Parallel Computer Architectures
Numerical Simulation in Oil Recovery
Atomic and Molecular Structure and Dynamics

D
A
B
C
E

CONTENTS

FOREWORD

This IMA Volume in Mathematics and its Applications

AMORPHOUS POLYMERS AND NON-NEWTONIAN FLUIDS

is in part the proceedings of a workshop which was an integral part of the 1984-85 IMA program on CONTINUUM PHYSICS AND PARTIAL DIFFERENTIAL EQUATIONS We are grateful to the Scientific Committee:

Haim Brezis
Constantine Dafermos
Jerry Ericksen
David Kinderlehrer

for planning and implementing an exciting and stimulating year-long program. We especially thank the Program Organizers, Jerry Ericksen, David Kinderlehrer, Stephen Prager and Matthew Tirrell for organizing a workshop which brought together scientists and mathematicians in a variety of areas for a fruitful exchange of ideas.

George R. Sell
Hans Weinberger

Preface

Experiences with amorphous polymers have supplied much of the motivation for developing novel kinds of molecular theory, to try to deal with the more significant features of systems involving very large molecules with many degrees of freedom. Similarly, the observations of many unusual macroscopic phenomena has stimulated efforts to develop linear and nonlinear theories of viscoelasticity to describe them. In either event, we are confronted not with a well-established, specific set of equations, but with a variety of equations, conforming to a loose pattern and suggested by general kinds of reasoning.

One challenge is to devise techniques for finding equations capable of delivering definite and reliable predictions. Related to this is the issue of discovering ways to better grasp the nature of solutions of those equations showing some promise.

This and the preceding volumes 2 and 4 provide different perspectives of some of the dynamical studies considered during the 1984-1985 I. M. A. Program, Continuum Physics and Partial Differential Equations. Contents of the other volumes, which may be found at the end of this book, contain other relevant titles.

This workshop brought together research workers in chemistry, engineering, mathematics, and physics from laboratory, industrial, and academic environments. The editors greatly appreciate the concerted efforts of the speakers and discussants to make their presentations intelligible to a mixed audience. Especial thanks are due to Gianfranco Capriz, Harry Frisch, M. Muthukumar, William Graessley, Stephen Prager, Matthew Tirrell, and Louis Zapas for their informal contributions.

The conference committee would like to take this opportunity to extend its gratitude to the staff of the I.M.A., Professors Weinberger and Sell, Mrs. Pat Kurth, and Mr. Robert Copeland for their assistance in arranging the workshop. Special thanks are due to Mrs. Debbie Bradley, Mrs. Patricia Brick, and Mrs. Kaye Smith for their preparation of the manuscripts. The workshop was supported by a grant from the Division of Industrial-Academic Cooperation of the Materials

Research Division as well as the Mathematical Sciences Division of the National Science Foundation. We gratefully acknowledge this support.

C. Dafermos
J. L. Ericksen
D. Kinderlehrer
Editors

MATHEMATICAL PROBLEMS IN THE KINETIC THEORY OF POLYMERIC FLUIDS

R. Byron Bird

Chemical Engineering Department and
Rheology Research Center
University of Wisconsin-Madison
Madison, Wisconsin 53706

An introduction to the kinetic theory of polymeric fluids is given, particularly that part of the theory that is of interest in polymer rheology and polymer fluid dynamics. The two main results of the kinetic theory are (a) a partial differential equation for the configurational distribution function of a polymer molecule in the liquid, and (b) an expression for the stress tensor involving averages calculated using the configurational distribution function. The principal mathematical challenge in this subject is that of solving the partial differential equation for the configurational distribution function. Several examples are given of complete and partial solutions to these equations.

1. Introduction

For incompressible fluids the laws of conservation of mass and momentum lead to the equations of continuity and motion:

$$\underline{\text{Continuity}}: \quad (\nabla \cdot \underline{v}) = 0 \tag{1.1}$$

$$\underline{\text{Motion}}: \quad \rho D\underline{v}/Dt = -[\nabla \cdot \underline{\underline{\pi}}] + \rho \underline{g} \tag{1.2}$$

Here ρ is the fluid density, $\underline{v}(\underline{r},t)$ the velocity field, and $\underline{g}$ the gravitational acceleration. It is convenient to write $\underline{\underline{\pi}}$ as the sum of two terms:

$$\underline{\underline{\pi}} = p\underline{\underline{\delta}} + \underline{\underline{\tau}} \tag{1.3}$$

in which p is a pressure (not uniquely determined for incompressible fluids) and $\underline{\underline{\tau}}$ is the "extra stress tensor", which is zero at equilibrium; $\underline{\underline{\delta}}$ is the unit tensor.

The flow of gases and of liquids composed of small molecules can be described well by the Newtonian constitutive equation:

$$\underline{\underline{\tau}} = -\mu(\nabla\underline{v} + (\nabla\underline{v})^\dagger) \equiv -\mu\dot{\underline{\underline{\gamma}}} \tag{1.4}$$

in which $(\nabla\underline{v})^\dagger$ is the transpose of $\nabla\underline{v}$, and $\dot{\underline{\underline{\gamma}}}$ is a rate-of-strain tensor. For dilute monatomic gases the dependence of μ on temperature and molecular weight has been deduced by the elaborate kinetic theory of Chapman and Enskog, and for multicomponent gas mixtures by Curtiss and Hirschfelder.[1] In these theories the gas molecules are modeled as point masses, which attract one another at large separations and repel one another at small separations. The formal kinetic theory consists of establishing two important relations: (i) the Boltzmann integrodifferential equation for the distribution function $f(\underline{r},\underline{p},t)$ in the phase space of a single gas molecule, and (ii) the expression for $\underline{\underline{\tau}}$, which gives the molecular momentum flux as an average value calculated using $f(\underline{r},\underline{p},t)$. The crucial mathematical problem is the solution of the integrodifferential equation for the distribution function.

Experiments have clearly shown that polymeric liquids (polymer solutions and undiluted polymers) are not described by Eq. 1.4. In the past several decades considerable effort has been expended to obtain the constitutive equation for these structurally complex liquids. In broad outline, the procedure is similar to that used in gas kinetic theory: a simple mechanical model for the macromolecule is selected; then the two basic relations are established via physical arguments (an equation for a distribution function and an expression for $\underline{\underline{\tau}}$); and finally the mathematical problem of solving the equation for the distribution function has to be attacked. This latter problem is extremely difficult, since for polymers, which have many internal degrees of freedom, the differential equations for the distribution functions inevitably lead to partial differential equations with many variables.

Most of the early workers in polymer kinetic theory did not attempt to derive a constitutive equation, but were content to extract only limited information such as the zero-shear-rate viscosity or the linear viscoelastic behavior. Notable exceptions were: Green and Tobolsky,[2] Lodge,[3] and Yamamoto,[4] who developed network theories; Rivlin,[5] who examined a dilute solution of freely jointed bead-rod chains; and Prager[6] and Giesekus,[7] who studied dilute solutions of dumbbells. In the last two decades the polymer-kinetic-theory field has grown, and now the most successful constitutive equations used in rheology and polymer fluid dynamics are those derived from some kind of kinetic-theory arguments. Here we give a brief introduction to the kinetic theory of polymeric liquids, in order to show the kinds of differential equations that arise for the configuration-space distribution functions.

2. Mechanical Models for Chainlike Polymer Macromolecules

In the kinetic theory the most widely studied polymer molecules are those in which monomeric units are joined end to end, to create a long chain-like chemical structure containing many thousands, or even millions, of atoms. These giant molecules are moderately flexible and, because of thermal motion, they are continually going through a wide range of configurations; in a flow field they may be oriented, deformed, and stretched out. Since the molecular architecture is well known, it is possible to use mechanical models capable of describing all the bond angles, bond distances, steric hindrances, and excluded volume to study the statistical properties of an isolated polymer molecule in a solution at equilibrium, and Flory's treatise[8] is devoted to just this subject. But for the much more complicated problem of using kinetic theory to describe nonequilibrium situations (that is, rheological and fluid dynamical phenomena), such detailed modeling is out of the question; it may be that Brownian dynamics techniques[9] will ultimately allow more faithful mechanical models to be used.

Up to the present time attention has been concentrated on hypothetical models made up of a series of spheres ("beads") joined linearly by some kind of connectors (usually "springs" or "rods"), with universal joints at the beads. The solvent is traditionally regarded as a continuum with Newtonian viscosity η_s, and the beads experience some kind of Stokes' law resistance as they move through the solvent. For the most commonly used models see Figure 1. For the sake of simplicity we consider here primarily chains with just two beads (that is, "dumbbells").

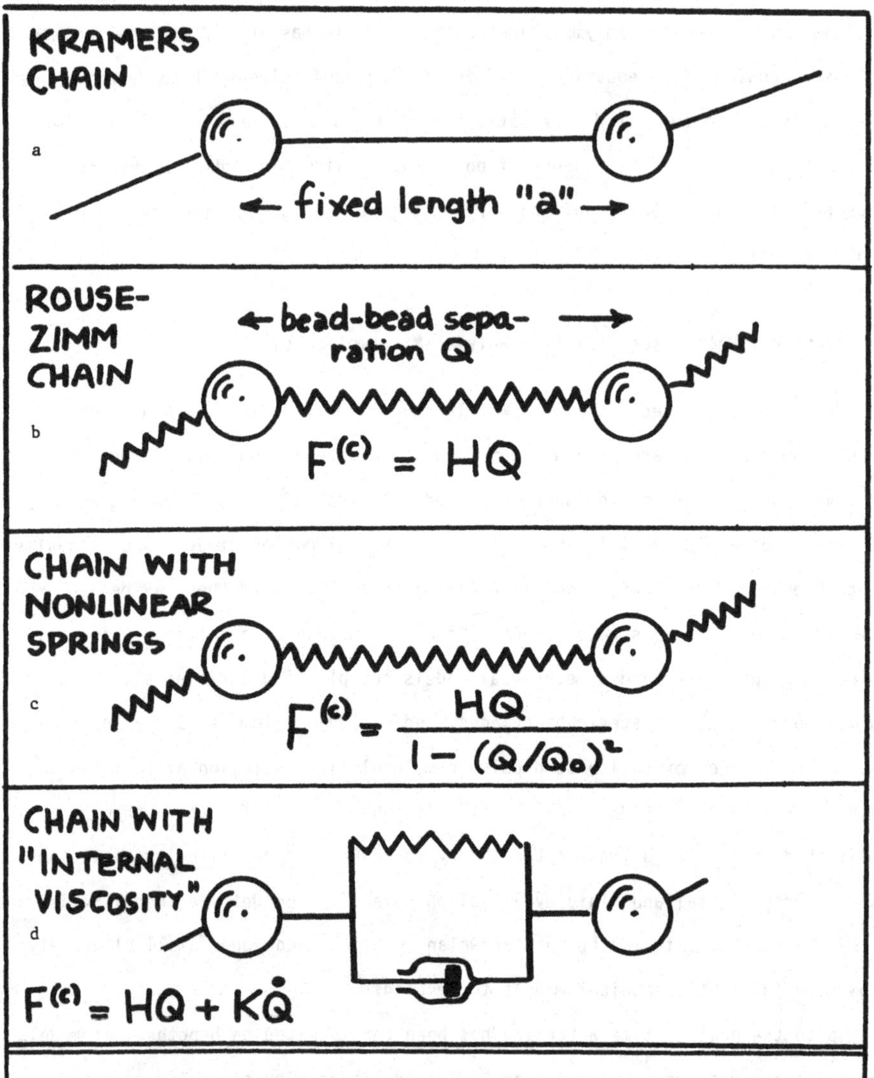

Fig. 1. Sketch showing several kinds of connectors used in chainlike models for polymers: (a) rigid rod; (b) linear (Hookean) spring; (c) finitely extensible nonlinear elastic (FENE) spring; (d) spring-dashpot combination describing "internal viscosity".

3. The Partial Differential Equation for the Distribution Function (Elastic-Dumbbell Models)[10]

As mentioned above, the kinetic theory consists of two main parts, one of which is the development of an equation for the configurational distribution function (sometimes called the "diffusion equation"). This "diffusion equation" is obtained by combining the equations of motion of the beads with an equation of continuity that describes the conservation of system points in the configuration space.

(a) The Equations of Motion for the Beads of the Dumbbell

We can write an equation of motion for each bead of the dumbbell indicating that the mass of the bead times its acceleration is equal to the sum of all forces acting on the bead. When we neglect the inertial terms containing the bead masses, we get a "force balance" among the various forces:

$$\underline{F}_\nu^{(h)} + \underline{F}_\nu^{(b)} + \underline{F}_\nu^{(\phi)} = \underline{0} \qquad (\nu = 1,2) \tag{3.1}$$

in which

$$\underline{F}_\nu^{(h)} = -[\underline{\underline{\zeta}} \cdot ([\![\dot{\underline{r}}_\nu]\!] - (\underline{v}_\nu + \underline{v}'_\nu))] \tag{3.2}$$

$$\underline{F}_\nu^{(b)} = -\frac{1}{\Psi}\left[\frac{\partial}{\partial \underline{r}_\nu} \cdot ([\![m(\dot{\underline{r}}_\nu - \underline{v})(\dot{\underline{r}}_\nu - \underline{v})]\!]\Psi)\right] \tag{3.3}$$

$$\underline{F}_\nu^{(\phi)} = -\frac{\partial}{\partial \underline{r}_\nu}\phi \tag{3.4}$$

In these equations $\underline{r}_\nu$ is the location of the νth bead, m is the mass of a bead, ϕ is the potential energy associated with the spring, and $\Psi(\underline{r}_1, \underline{r}_2, t)$ is the configurational distribution function for the dumbbell. The double brackets $[\![\ ]\!]$ indicate an average in the velocity space. We now discuss each of the contributions to the force balance.

Equation 3.2 describes the hydrodynamic force acting on bead ν. According to this expression the force is proportional to the difference between the bead velocity $\dot{\underline{r}}_\nu$ (appropriately averaged with respect to the velocity distribution) and the velocity $(\underline{v}_\nu + \underline{v}'_\nu)$ of the solution at bead ν. The velocity $\underline{v}_\nu = \underline{v}_0 + [\underline{\underline{\kappa}} \cdot \underline{r}_\nu]$ is the imposed homogeneous flow field at bead ν (here $\underline{\underline{\kappa}}(t)$ and $\underline{v}_0(t)$ are position independent), and $\underline{v}'_\nu$ is the perturbation of the flow field at bead ν resulting from the motion of the other bead; this perturbation is referred to as "hydrodynamic interaction". In this discussion we neglect hydrodynamic interaction and $\underline{v}'_\nu$. We note further that, according to Eq. 3.2, the hydrodynamic drag force is not necessarily collinear with the velocity difference since the coefficient of proportionality is a symmetric second-order tensor $\underline{\underline{\zeta}}$, called the "friction tensor". In what follows we take this tensor to be isotropic, so that $\underline{\underline{\zeta}} = \zeta\underline{\underline{\delta}}$, where the scalar ζ is called the "friction coefficient". The beads actually execute very tortuous paths as they move about in the solvent, but by using the velocity-space average of $\dot{\underline{r}}_\nu$ we obtain a kind of "smoothed out" drag force.

Equation 3.3 is a smoothed-out Brownian motion force. The true Brownian motion force would be a rapidly and irregularly fluctuating function. Instead of the latter we use a statistically averaged force,[11,12] the origin of which can be understood from a complete phase-space kinetic theory.[13] It should be noted that the expression for the Brownian force has the form of the divergence of a momentum flux with respect to the solution velocity, $\underline{v}$, at the center of mass of the dumbbell. In almost all kinetic theories published so far, equilibration in momentum space has been tacitly assumed; that is, the velocity distribution is Maxwellian. When this is done the Brownian force assumed the much simpler form $\underline{F}_\nu^{(b)} = -kT(\partial \ln \Psi / \partial \underline{r}_\nu)$.

Equation 3.4 gives the force $\underline{F}_\nu^{(\phi)}$ on the νth bead resulting from the intramolecular potential energy. For the simple model under consideration this is just the force acting through the spring in the dumbbell. For the dumbbell models the forces on the two beads are equal and opposite, so that it is useful to define a "connector force" $\underline{F}^{(c)}$ by $\underline{F}^{(c)} = \underline{F}_1^{(\phi)} = -\underline{F}_2^{(\phi)}$; see Fig. 1(b,c) for two examples of connector-force expressions.

We now assume that the friction tensor is a multiple of the unit tensor, that the Maxwellian velocity distribution is used in the Brownian motion term, and that hydrodynamic interaction is neglected; then the equations of motion become:

$$-\zeta([\dot{\underline{r}}_\nu] - \underline{v}_0 - [\underline{\underline{\kappa}} \cdot \underline{r}_\nu]) - kT \frac{\partial}{\partial \underline{r}_\nu} \ln \Psi + \underline{F}_\nu^{(\phi)} = \underline{0} \quad (\nu = 1,2) \tag{3.5}$$

When these two equations are added together and then divided by 2, we get the equation of motion for the center of mass $\underline{r}_c = (1/2)(\underline{r}_1 + \underline{r}_2)$; when they are subtracted, we get the equation of motion for the dumbbell connector vector $\underline{Q} = \underline{r}_2 - \underline{r}_1$:

$$[\dot{\underline{r}}_c] = \underline{v}_0 + [\underline{\underline{\kappa}} \cdot \underline{r}_c] \tag{3.6}$$

$$[\dot{\underline{Q}}] = [\underline{\underline{\kappa}} \cdot \underline{Q}] - \frac{2kT}{\zeta} \frac{\partial}{\partial \underline{Q}} \ln \psi - \frac{2}{\zeta} \underline{F}^{(c)} \tag{3.7}$$

Here we have introduced $\psi(\underline{Q},t)$ which is related to $\Psi(\underline{r}_1,\underline{r}_2,t)$ by $\Psi = n\psi$, where n is the number density of dumbbells. The first of the equations above shows that the center of mass of the dumbbell moves on the average with the solution velocity at the location of the center of mass. The second equation is used presently to obtain the diffusion equation for $\psi(\underline{Q},t)$.

(b) <u>The Equation of Continuity for</u> $\psi(Q,t)$

The distribution function $\Psi(\underline{r}_1,\underline{r}_2,t)$ must satisfy a continuity equation:

$$\frac{\partial \Psi}{\partial t} = -\left(\frac{\partial}{\partial \underline{r}_1} \cdot [\dot{\underline{r}}_1]\Psi\right) - \left(\frac{\partial}{\partial \underline{r}_2} \cdot [\dot{\underline{r}}_2]\Psi\right) \tag{3.8}$$

which accounts for conservation of system points in the six-dimensional configuration space.

By a change of variables this can be rewritten for $\psi(\underline{Q},t)$ as:

$$\frac{\partial \psi}{\partial t} = -\left(\frac{\partial}{\partial \underline{Q}} \cdot [\dot{\underline{Q}}]\psi\right) \tag{3.9}$$

This is the continuity equation in the three-dimensional internal configuration space.

(c) **The "Diffusion Equation" for $\psi(\underline{Q},t)$**

Substitution of $[\dot{\underline{Q}}]$ from Eq. 3.7 into Eq. 3.9 gives the "diffusion equation":

$$\frac{\partial\psi}{\partial t} = -\left(\frac{\partial}{\partial \underline{Q}} \cdot \{[\underline{\underline{\kappa}} \cdot \underline{Q}]\psi - \frac{2kT}{\zeta}\frac{\partial}{\partial \underline{Q}}\psi - \frac{2}{\zeta}\underline{F}^{(c)}\psi\}\right) \tag{3.10}$$

This is the second-order partial differential equation that describes the way in which the distribution of configurations changes with time when the time-dependent homogeneous velocity field is described by $\underline{\underline{\kappa}}(t)$ and the dumbbell spring force is given as $\underline{F}^{(c)}$. This is the fundamental differential equation in the elementary elastic-dumbbell kinetic theory. More complicated differential equations result if the friction tensor $\underline{\underline{\zeta}}$ is used,[11,14] or if hydrodynamic interaction is used,[10] or if the velocity distribution is not Maxwellian.[11,12,14] And, of course, if the theory is extended to chains, the analogs of Eq. 3.10 are equations in the (3N - 3)-dimensional chain configuration space, where N is the number of beads (see Ref. 10, Chapter 12).

4. The Stress-Tensor Expression (The Elastic-Dumbbell Model)[10,14]

In an elastic dumbbell solution there are three contributions to the stress tensor: the solvent contribution; a contribution from the tensions in the springs, $\underline{F}^{(c)}$; and a contribution from the bead momentum flux (the latter being reminiscent of the dilute-gas stress-tensor expression):

$$\underline{\underline{\pi}} = \underline{\underline{\pi}}_s - n\,\langle \underline{Q}\underline{F}^{(c)}\rangle + nm\sum_{\nu=1}^{2}\langle(\dot{\underline{r}}_\nu - \underline{v})(\dot{\underline{r}}_\nu - \underline{v})\rangle \tag{4.1}$$

Here < > stands for an average in the phase space of the dumbbell. The term containing $\underline{F}^{(c)}$ is usually the most important of the three terms. The close connection between the momentum-flux contribution to the stress tensor and $\underline{F}_\nu^{(b)}$ in Eq. 3.3 has not been generally appreciated. Equation 4.1 is valid whether or not hydrodynamic interaction is included and whether or not the friction tensor ζ is isotropic; also no assumption has been introduced regarding the velocity distribution.

For a velocity distribution Maxwellian about $\underline{v}$ the last term in Eq. 4.1 becomes isotropic, and the expression for the extra stress tensor $\underline{\underline{\tau}}$ then becomes Eq. A in Table 1. Equivalent expressions for $\underline{\underline{\tau}}$ are given in Eqs. B and C in terms of $\underline{F}_{\nu}^{(\phi)}$ and $\underline{F}_{\nu}^{(h)}$ rather than $\underline{F}^{(c)}$. Equation D is a less general expression.

Table 1

Expressions for the Extra Stress Tensor for Elastic Dumbbells
(Maxwellian Velocity Distribution Assumed)

$$\underline{Q} = \underline{r}_2 - \underline{r}_1$$
$$\underline{R}_{\nu} = \underline{r}_{\nu} - \underline{r}_c$$
$$\underline{r}_c = (1/2)(\underline{r}_1 + \underline{r}_2)$$

Kramers[1]: $$\underline{\underline{\tau}} = -\eta_s \dot{\underline{\underline{\gamma}}} - n\langle \underline{Q}\underline{F}^{(c)}\rangle + nkT\underline{\underline{\delta}} \qquad \text{(A)}$$

Modified-Kramers: $$\underline{\underline{\tau}} = -\eta_s \dot{\underline{\underline{\gamma}}} + n \sum_{\nu} \langle \underline{R}_{\nu}\underline{F}_{\nu}^{(\phi)}\rangle + nkT\underline{\underline{\delta}} \qquad \text{(B)}$$

Kramers-Kirkwood[2]: $$\underline{\underline{\tau}} = -\eta_s \dot{\underline{\underline{\gamma}}} - n \sum_{\nu} \langle \underline{R}_{\nu}\underline{F}_{\nu}^{(h)}\rangle \qquad \text{(C)}$$

Giesekus[3]: $$\underline{\underline{\tau}} = -\eta_s \dot{\underline{\underline{\gamma}}} + \frac{n}{4} \zeta \langle \underline{Q}\underline{Q}\rangle_{(1)} \qquad \text{(D)}$$

[1] H.A. Kramers, Physica, 11, 1-19 (1944), in Dutch.

[2] J.G. Kirkwood, Macromolecules, Gordon and Breach, New York (1967).

[3] H. Giesekus, Rheol. Acta, 2, 50-62 (1962); this expression is less general than Eqs. A-C, in that the friction tensor must be isotropic and hydrodynamic interaction is neglected. The subscript (1) indicates a convected derivative defined in Eq. A.9.

5. Solutions to the "Diffusion Equation" (Elastic-Dumbbell Model)

We now return to Eq. 3.10 which we want to solve for any homogeneous flow field $\underline{\underline{\kappa}}(t)$. The following solutions are known:

a. **Steady, Homogeneous, Potential Flow with Any Kind of Spring** [10]

For this flow $\underline{\underline{\kappa}} = \underline{\underline{\kappa}}^{\dagger}$, and ψ is given by:

$$\psi(\underline{Q}) \sim e^{(\zeta/4kT)(\underline{\underline{\kappa}}:\underline{Q}\underline{Q})} e^{-\phi^{(c)}/kT} \tag{5.1}$$

in which $\phi^{(c)}$ is defined by $\underline{F}^{(c)} = +(\partial/\partial\underline{Q})\phi^{(c)}$. This solution includes the equilibrium distribution function $\psi_{eq}(\underline{Q})$ for which $\underline{\underline{\kappa}} = \underline{\underline{0}}$.

b. Arbitrary flow with Hookean dumbbells [15] ($\underline{F}^{(c)} = H\underline{Q}$)

For this choice of spring force-law:

$$\psi(\underline{Q},t) = \frac{(H/2\pi kT)^{3/2}}{\sqrt{\det \underline{\underline{\alpha}}}} e^{-(H/2kT)(\underline{\underline{\alpha}}^{-1}:\underline{Q}\underline{Q})} \tag{5.2}$$

in which

$$\underline{\underline{\alpha}}(t) = \underline{\underline{\delta}} - \frac{1}{\lambda_H}\int_{-\infty}^{t} e^{-(t-t')/\lambda_H} \underline{\underline{\gamma}}_{[0]}(t,t')dt' \tag{5.3}$$

Here $\lambda_H = \zeta/4H$ is a time constant, and $\underline{\underline{\gamma}}_{[0]}$ is the finite strain tensor of Eq. A.13. Note that $\underline{\underline{\kappa}}$ appears in the differential equation, but that $\underline{\underline{\gamma}}_{[0]}$ appears in the solution; the relation between $\underline{\underline{\gamma}}_{[0]}$ and $\underline{\underline{\kappa}}$ is far from simple (see Eqs. A.13 and A.16).

c. **Steady Slow Flows with FENE Dumbbells** [10,16]

A perturbation solution valid for small velocity gradients is:

$$\psi \sim [1 - (\frac{Q}{Q_0})^2]^{b/2}$$

$$\Big(\cdot (1 + (\frac{H\lambda_H}{2kT})(\dot{\underline{\underline{\gamma}}}:\underline{Q}\underline{Q}) + (\frac{H\lambda_H}{2kT})^2 [\frac{Q_0^{\,4}}{(b+5)(b+7)}(\dot{\underline{\underline{\gamma}}}:\dot{\underline{\underline{\gamma}}})$$

$$+ \frac{4Q_0^{\,2}}{(2b+7)}(1 - \frac{1}{2}(\frac{Q}{Q_0})^2)(\{\underline{\underline{\gamma}}\cdot\underline{\underline{\omega}}\}:\underline{Q}\underline{Q}) + \frac{1}{2}(\underline{\underline{\gamma}}:\underline{Q}\underline{Q})^2] + \ldots\Big) \tag{5.4}$$

in which $\lambda_H = \zeta/4H$ and $b = HQ_0^{\,2}/kT$; when $b \to \infty$ the slow-flow expansion for Hookean dumbbells is obtained.

d. **Steady Shear Flow with FENE Dumbbells**

A numerical solution of the diffusion equation has been obtained by a Galerkin procedure, both for ζ = constant[17] and for ζ a linear function of the interbead separation.[18]

These are the only solutions of the diffusion equation for elastic dumbbells that we know of for isotropic friction tensors. There is a technique for getting the complete constitutive equation for Hookean dumbbells without actually finding Eqs. 5.2 and 5.3; that is, the problem of solving the diffusion equation can be circumvented but only for this one special model. A similar technique has been used to get approximate constitutive equations for FENE dumbbells for both isotropic[17,19] and nonisotropic[11] friction tensors.

6. Solutions of the "Diffusion Equation" (Rigid Dumbbell Model)[10,20]

For a dilute solution of rigid rodlike macromolecules we can use the rigid dumbbell model (see Fig. 1(a)). Rigid dumbbell results may be extended to the multibead dumbbell, and hydrodynamic interaction can be included.[20]

For the rigid dumbbell the equation for the distribution function $f(\underline{u},t) = \Psi(\theta,\phi,t)/n \sin\theta$ can be shown to be

$$\frac{\partial f}{\partial t} = \frac{1}{6\lambda}\left(\frac{\partial}{\partial \underline{u}}\cdot\frac{\partial}{\partial \underline{u}} f\right) - \left(\frac{\partial}{\partial \underline{u}}\cdot[\underline{\underline{\kappa}}\cdot\underline{u} - \underline{\underline{\kappa}}:\underline{uuu}]f\right) \tag{6.1}$$

in which $\lambda = \zeta a^2/12kT$ is the time constant, $\underline{u}$ is a unit vector in the direction of the rod, and $\partial/\partial\underline{u} = r[(\underline{\underline{\delta}} - \underline{uu})\cdot\nabla]$ is the gradient operator on the surface of a sphere of radius r.

The stress tensor expression is

$$\begin{aligned}\underline{\underline{\tau}} &= -\eta_s\dot{\underline{\underline{\gamma}}} - 3nkT[\langle\underline{uu} - \tfrac{1}{3}\underline{\underline{\delta}}\rangle + 2\lambda\underline{\underline{\kappa}}:\langle\underline{uuuu}\rangle] \\ &= -\eta_s\dot{\underline{\underline{\gamma}}} + 3nkT\lambda\langle\underline{uu}\rangle_{(1)}\end{aligned} \tag{6.2}$$

the two different forms corresponding to Eqs. A and D of Table 1. Equations 6.1 and 6.2 are valid if there is equilibration in momentum space.

No general solution (for all $\underline{\underline{\kappa}}(t)$) is known, but the following partial solutions are available:

a. **Steady, Homogeneous, Potential Flow:**

For this flow[10] we have:

$$\psi \sim e^{3\lambda(\underline{\underline{\kappa}}:\underline{uu})} \tag{6.3}$$

in which $\underline{\underline{\kappa}}$ is symmetric.

b. **Arbitrary Flows**[10]

A memory-integral expansion for f has been found:

$$f = \frac{1}{4\pi}\left(1 + (6\lambda)\phi_1 + (6\lambda)^2\phi_2 + \ldots\right) \tag{6.4}$$

where

$$\phi_1 = \frac{1}{4\lambda}\int_{-\infty}^{t} e^{-(t-t')/\lambda}\,(\dot{\underline{\underline{\gamma}}}':\underline{uu})dt'$$

$$\phi_2 = \frac{1}{\lambda^2}\int_{-\infty}^{t}\int_{-\infty}^{t'}\left(-\frac{1}{168}(\dot{\underline{\underline{\gamma}}}':\dot{\underline{\underline{\gamma}}}'') + \frac{1}{56}(\{\dot{\underline{\underline{\gamma}}}'\cdot\dot{\underline{\underline{\gamma}}}''\}:\underline{uu})\right) e^{-(t-t'')/\lambda}\,dt''dt'$$

$$+ \frac{1}{\lambda^2}\int_{-\infty}^{t}\int_{-\infty}^{t'}\Big(+\frac{1}{168}(\dot{\underline{\underline{\gamma}}}':\dot{\underline{\underline{\gamma}}}'') - \frac{5}{84}(\{\dot{\underline{\underline{\gamma}}}'\cdot\dot{\underline{\underline{\gamma}}}''\}:\underline{uu})$$

$$+ \frac{5}{48}(\dot{\underline{\underline{\gamma}}}':\underline{uu})(\dot{\underline{\underline{\gamma}}}'':\underline{uu})\Big] e^{-(7/3)(t-t')/\lambda}\, e^{-(t-t'')/\lambda}\,dt''dt' \tag{6.6}$$

In these equations $\dot{\underline{\underline{\gamma}}}'$ stands for $\dot{\underline{\underline{\gamma}}}(t')$.

c. **Steady Shear Flow** [10,21]

An extensive numerical solution has been obtained by expanding the solution in a finite series of spherical harmonics.

To our knowledge no one has sought a solution in terms of the finite strain tensor $\underline{\underline{\gamma}}_{[0]}$ instead of in terms of $\dot{\underline{\underline{\gamma}}}$ or $\underline{\underline{\kappa}}$.

7. Kinetic Theory of Polymer Melts

Up to this point we have discussed molecular theories of dilute solutions. In recent years attention has shifted to the "reptation theories" of undiluted polymers as a result of the pioneering papers by Doi and Edwards.[22] An alternative phase-space kinetic theory has also been proposed,[14,23] which is based on the Curtiss-Bird-Hassager formulation for polymeric systems.[24] A comparison of the two approaches (with their different molecular models, theoretical basis, and assumptions) has been given.[25]

In both theories the equation for the distribution function is given by[22,23]

$$\frac{\partial f}{\partial t} = \frac{1}{\lambda}\frac{\partial^2 f}{\partial \sigma^2} - \left(\frac{\partial}{\partial \underline{u}} \cdot [\kappa \cdot u - \kappa : uuu]f\right) \qquad (7.1)$$

in which $\lambda = N^{3+\beta}\zeta a^2/2kT$ is the time constant, and σ is the fractional distance along a polymer chain regarded as a "smoothed-out" Kramers freely jointed bead-rod chain of N beads and $N - 1$ rigid links with link length a; the parameter β is about 0.4. The distribution function $f(\underline{u},\sigma,t)$ gives the probability that, at a distance σ along the chain, the link orientation is given by a unit vector $\underline{u}$ at time t. The last term in Eq. 7.1, describing the twisting of a link by the flow field, is identical to the last term in Eq. 6.1.

The stress tensor expression is given (in the Curtiss-Bird theory[23]) by:

$$\underline{\underline{\tau}} = -NnkT\Big[\int_0^1 \langle \underline{uu} - \tfrac{1}{3}\underline{\underline{\delta}} \rangle\, d\sigma + \varepsilon\lambda\underline{\underline{\kappa}} : \int_0^1 \sigma(1 - \sigma)\langle \underline{uuuu} \rangle d\sigma\Big] \qquad (7.2)$$

in which ε is a parameter, whose value is found experimentally to be in the range 0.3 - 0.5 for many fluids.[23] The ε-parameter is introduced to describe the non-isotropic Stokes drag on the beads of the Kramers chain ($\varepsilon = 1$ corresponds to isotropic drag, and $\varepsilon = 0$ corresponds to the Doi-Edwards theory).

A complete constitutive equation can be obtained by solving Eq. 7.1 for $f(\underline{u},\sigma,t)$ and using this function to calculate the $\langle\ \rangle$-averages in Eq. 7.2. Doi and Edwards[22] solved Eq. 7.1, but a much simpler solution has been found:[23]

$$f(\underline{u},\sigma,t) = \frac{1}{\lambda}\int_{-\infty}^{t} F(\sigma,t-t')[1 + (\underline{\underline{\gamma}}^{[0]} : \underline{uu})]^{-3/2}\, dt' \qquad (7.3)$$

in which

$$F(\sigma,s) = \sum_{\alpha,\mathrm{odd}} \alpha(\sin \pi\alpha\sigma) \exp(-\pi^2\alpha^2 s/\lambda) \tag{7.4}$$

Note that Eq. 7.1 contains $\underline{\underline{\kappa}} = (\nabla \underline{v})^\dagger$ as the kinematic tensor, whereas Eq. 7.3 contains the finite strain tensor $\underline{\underline{\gamma}}^{[0]}$; according to Eqs. A.12 and A.15, these two kinematic tensors are not simply related (cf. the comments following Eqs. 5.2 and 5.3).

8. Conclusions

Evidence is accumulating that the bead-spring-rod models of macromolecules along with the kinetic theories can describe a substantial body of rheological data. The kinetic theories also lead to constitutive equations that are useful for polymer-fluid-dynamics problem solving. A major impediment to progress is the paucity of information about methods for solving the "diffusion equation" for the configurational distribution function. Analytical solutions are needed most of all, but asymptotic, approximate, and numerical solutions are also valuable. New mathematical procedures are needed particularly for chain molecules with many internal coordinates.

Acknowledgements

The author is indebted to the National Science Foundation (Grant CPE-8104705), the Vilas Trust Fund of the University of Wisconsin, and a MacArthur Professorship for financial support in the field of polymer fluid dynamics.

APPENDIX A

CONTINUUM MECHANICS SUMMARY[26]

1. Description of Flow Fields

The motion of a fluid may be described in two ways: (a) in terms of the fluid velocity field, or (b) in terms of the fluid particle trajectories. Both descriptions are used in the definitions of kinematic tensors.

a. The fluid velocity field and closely related quantities

The velocity of a fluid at position $\underline{r}$ at time t is denoted by $\underline{v}(\underline{r},t)$. A flow is said to be homogeneous if the velocity gradients are the same at all points in the fluid so that,

$$\underline{v} = \underline{v}_0 + [\underline{\underline{\kappa}}(t) \cdot \underline{r}] \tag{A.1}$$

where $\underline{\underline{\kappa}}$ and $\underline{v}_0$ are independent of position but may be functions of time. For incompressible fluids $\mathrm{tr}\,\underline{\underline{\kappa}} = 0$.

For any velocity field $\underline{v}$, the rate-of-strain tensor (or rate-of-deformation tensor) $\dot{\underline{\underline{\gamma}}}$ is defined by:

$$\dot{\underline{\underline{\gamma}}} = \underline{\nabla v} + (\underline{\nabla v})^{\dagger} = \underline{\underline{\kappa}}^{\dagger} + \underline{\underline{\kappa}} \tag{A.2}$$

where $\dagger$ indicates the transpose of the tensor.

The vorticity tensor is defined by:

$$\underline{\underline{\omega}} = \underline{\nabla v} - (\underline{\nabla v})^{\dagger} = \underline{\underline{\kappa}}^{\dagger} - \underline{\underline{\kappa}} \tag{A.3}$$

b. The fluid particle trajectories and closely related quantities

Let a fluid particle be at position $\underline{r}$ at time t and at a position $\underline{r}'$ at some other time t'. Then the motion of the fluid particle is given by the displacment functions

$$\underline{r} = \underline{r}(\underline{r}',t',t) \qquad \text{or} \qquad \underline{r}' = \underline{r}'(\underline{r},t,t') \tag{A.4}$$

Specification of the displacement functions for all fluid particles is equivalent to giving the velocity field for all positions in space and all times.

The Cartesian components of the displacement gradient tensors $\underline{\underline{\Delta}}$ and $\underline{\underline{E}}$ are given by:

$$\Delta_{ij}(\underline{r},t,t') = \partial x_i'/\partial x_j \; ; \quad E_{ij}(\underline{r},t,t') = \partial x_i/\partial x_j' \tag{A.5}$$

The tensors $\underline{\underline{\Delta}}$ and $\underline{\underline{E}}$ are inverse to one another, and are both equal to the unit tensor when $t' = t$. For incompressible fluids $\det \underline{\underline{\Delta}} = 1$ and $\det \underline{\underline{E}} = 1$.

2. Time Derivatives

Four different kinds of time derivatives are used; we illustrate these in terms of derivatives of a second-order tensor $\underline{\underline{\Lambda}}$:

a. The partial derivative $\partial\underline{\underline{\Lambda}}/\partial t$ is a derivative with respect to time, the position is space being held fixed.

b. The substantial (or material) derivative gives the time rate of change following a fluid particle:

$$\frac{D}{Dt}\underline{\underline{\Lambda}} = \frac{\partial}{\partial t}\underline{\underline{\Lambda}} + \{\underline{v} \cdot \nabla\underline{\underline{\Lambda}}\} \tag{A.6}$$

c. The (Jaumann) corotational derivative gives the time rate of change, following a particle, in a coordinate frame rotating with the instantneous fluid angular velocity at the particle:

$$\frac{\mathcal{D}}{\mathcal{D}t}\underline{\underline{\Lambda}} = \frac{D}{Dt}\underline{\underline{\Lambda}} + \frac{1}{2}\{\underline{\underline{\omega}} \cdot \underline{\underline{\Lambda}} - \underline{\underline{\Lambda}} \cdot \underline{\underline{\omega}}\} \tag{A.7}$$

d. The (Oldroyd) convected derivatives (or codeformational derivatives)[26] give the time rate of change, following the particle, in a local coordinate system embedded in the fluid and moving along with it. Two such derivatives can be obtained, $\underline{\underline{\Lambda}}^{(1)}$ and $\underline{\underline{\Lambda}}_{(1)}$:

$$\underline{\underline{\Lambda}}^{(1)} = \frac{\mathcal{D}}{\mathcal{D}t}\underline{\underline{\Lambda}} + \frac{1}{2}\{\dot{\underline{\underline{\gamma}}} \cdot \underline{\underline{\Lambda}} + \underline{\underline{\Lambda}} \cdot \dot{\underline{\underline{\gamma}}}\} \tag{A.8}$$

$$\underline{\underline{\Lambda}}_{(1)} = \frac{\mathcal{D}}{\mathcal{D}t}\underline{\underline{\Lambda}} - \frac{1}{2}\{\dot{\underline{\underline{\gamma}}} \cdot \underline{\underline{\Lambda}} + \underline{\underline{\Lambda}} \cdot \dot{\underline{\underline{\gamma}}}\} \tag{A.9}$$

Higher-order convected derivatives are defined as $\underline{\underline{\Delta}}^{(2)} = (\underline{\underline{\Delta}}^{(1)})^{(1)}$, $\underline{\underline{\Delta}}_{(2)} = (\underline{\underline{\Delta}}_{(1)})_{(1)}$, etc.

3. Kinematic Tensors

Only the most commonly encountered kinematic tensors are summarized here.

a. Higher order rate-of-strain tensors (obtained from the velocity field)

The rate of strain tensor $\dot{\underline{\underline{\gamma}}}$ is defined in Eq. 2. When used in equations along with its convected derivatives we use the notation $\underline{\underline{\gamma}}^{(1)}$ or $\underline{\underline{\gamma}}_{(1)}$ for this quantity; but we must emphasize that the index "(1)" in this case does not refer to the first convected derivative of the infinitesimal strain tensor $\underline{\underline{\gamma}}$. The higher-order convected derivatives are called the Oldroyd[26] second, third, etc., rate-of-strain tensors:

$$\underline{\underline{\gamma}}^{(2)} = (\underline{\underline{\gamma}}^{(1)})^{(1)}, \quad \underline{\underline{\gamma}}^{(3)} = (\underline{\underline{\gamma}}^{(2)})^{(1)}, \text{ etc.} \tag{A.10}$$

$$\underline{\underline{\gamma}}_{(2)} = (\underline{\underline{\gamma}}_{(2)})_{(1)}, \quad \underline{\underline{\gamma}}_{(3)} = (\underline{\underline{\gamma}}_{(2)})_{(1)}, \text{ etc.} \tag{A.11}$$

The $\underline{\underline{\gamma}}^{(n)}$ are sometimes called the Rivlin-Ericksen tensors.

b. Finite strain tensors (obtained from the displacement functions)

We use two relative finite strain tensors, $\underline{\underline{\gamma}}^{[0]}$ and $\underline{\underline{\gamma}}_{[0]}$:

$$\underline{\underline{\gamma}}^{[0]}(\underline{r},t,t') = \{\underline{\underline{\Delta}}^{\dagger} \cdot \underline{\underline{\Delta}} - \underline{\underline{\delta}}\} \tag{A.12}$$

$$\underline{\underline{\gamma}}_{[0]}(\underline{r},t,t') = \{\underline{\underline{\delta}} - \underline{\underline{E}} \cdot \underline{\underline{E}}^{\dagger}\} \tag{A.13}$$

in which $\underline{\underline{\Delta}}$ and $\underline{\underline{E}}$ are the tensors whose Cartesian components were given in Eq. 5. They both simplify to the infinitesimal strain tensor $\underline{\underline{\gamma}}$ for vanishingly small displacement gradients. Higher-order time derivatives are designated as $\underline{\underline{\gamma}}^{[n]}(\underline{r},t,t') = \partial^n \underline{\underline{\gamma}}^{[0]}/\partial t'^n$ and $\underline{\underline{\gamma}}_{[n]}(\underline{r},t,t') = \partial^n \underline{\underline{\gamma}}_{[0]}/\partial t'^n$. Furthermore,

$$\underline{\underline{\gamma}}^{[n]}(\underline{r},t,t) = \underline{\underline{\gamma}}^{(n)}(\underline{r},t); \quad \underline{\underline{\gamma}}_{[n]}(\underline{r},t,t) = \underline{\underline{\gamma}}_{(n)}(\underline{r},t) \tag{A.14}$$

gives the relation between the [ ] and ( ) quantities.

Finally, we note that the relation between the finite strain tensors ($\underline{\underline{\gamma}}^{[0]}$ and $\underline{\underline{\gamma}}_{[0]}$) and the transpose of the velocity-gradient tensor $\underline{\underline{\kappa}} = (\nabla\underline{v})^{\dagger}$ can be obtained from:

$$\underline{\underline{\Delta}} = \underline{\underline{\delta}} - \int_{t'}^{t} \underline{\underline{\kappa}}'' dt'' + \int_{t'}^{t} \int_{t''}^{t} \{\underline{\underline{\kappa}}' \cdot \underline{\underline{\kappa}}''\} dt''' dt'' - \ldots \tag{A.15}$$

$$\underline{\underline{E}} = \underline{\underline{\delta}} + \int_{t'}^{t} \underline{\underline{\kappa}}'' dt'' + \int_{t'}^{t} \int_{t''}^{t} \{\underline{\underline{\kappa}}'' \cdot \underline{\underline{\kappa}}''\} dt''' dt'' + \ldots \tag{A.16}$$

along with Eqs. A.12 and 13; in these equations $\underline{\underline{\Delta}}$ and $\underline{\underline{E}}$ are functions of the "particle label" ($\underline{r}$,t) and t'; and $\underline{\underline{\kappa}}''$ is an abbreviation for $\underline{\underline{\kappa}}(\underline{r},t,t'')$.

References

1. For a full discussion and references to the original literature, see J.O. Hirschfelder, C.F. Curtiss, and R.B. Bird, Molecular Theory of Gases and Liquids, Wiley, New York, Second Corrected Printing (1964).

2. M.S. Green and A.V. Tobolsky, J. Chem. Phys., 14, 80-92 (1946).

3. A.S. Lodge, Trans. Faraday Soc., 52, 120-130 (1956); Rheol. Acta, 7, 379-392 (1968).

4. M. Yamamoto, J. Phys. Soc. Japan, 11, 413-421 (1956); 12, 1148-1158 (1957), 13, 1200-1211 (1958).

5. R.S. Rivlin, Trans. Faraday Soc., 45, 739-748 (1949); see also O. Hassager, J. Chem. Phys., 60, 2111-2124 (1974).

6. S. Prager, Trans. Soc. Rheol., 1, 53-62 (1957).

7. H. Giesekus, Kolloid-Z., 147-149, 29-45 (1956).

8. P.J. Flory, Statistical Mechanics of Chain Molecules, Wiley-Interscience, New York (1969).

9. P.J. Dotson, J. Chem. Phys., 79, 5730-5731 (1983); Ph.D. Thesis, University of Wisconsin-Madison (1984).

10. R.B. Bird, O. Hassager, R.C. Armstrong, and C.F. Curtiss, Dynamics of Polymeric Liquids, Vol. 2, Kinetic Theory, Wiley, New York (1977).

11. R.B. Bird and J.R. DeAguiar, J. Non-Newtonian Fluid Mech., 13, 149-160 (1983); the authors inferred Eq. 3.3 from Eq. 14.6-9 of Ref. 10.

12. R.B. Bird, X.J. Fan, and C.F. Curtiss, J. Non-Newtonian Fluid Mechanics, 15, 85-92 (1984).

13. C.F. Curtiss, Unpublished derivation; Curtiss's complete derivation will be given in Chapter 18 of the second edition of Ref. 10, which is now in press.

14. C.F. Curtiss and R.B. Bird, Physica, 118A, 191-204 (1983).

15. P.H. van Wiechen and H.C. Booij, J. Engr. Math., 5, 89-98 (1971), obtained the solution for a multibead (Rouse) chain, and A.S. Lodge and Y. Wu, Rheol. Acta, 10, 539-553 (1971), obtained a solution for the bead-spring chain with hydrodynamic interaction (the Zimm model).

16. R.C. Armstrong, J. Chem. Phys., 60, 724-728, 729-733 (1974).

17. X.J. Fan, J. Non-Newtonian Fluid Mech., 17, 125-144 (1985).

18. X.J. Fan, R.B. Bird and M. Renardy, J. Non-Newtonian Fluid Mech., 18, 255-272 (1985).

19. R.B. Bird, P.J. Dotson, and N.L. Johnson, J. Non-Newtonian Fluid Mech., 7, 213-235 (1980).

20. R.B. Bird and C.F. Curtiss, J. Non-Newtonian Fluid Mech., 14, 85-101 (1984).

21. W.E. Stewart and J.P. Sørensen, Trans. Soc. Rheol., 16, 1-13 (1972).

22. M. Doi and S.F. Edwards, J. Chem. Soc., Faraday Trans. II, 74, 1789-1832 (1978); 75, 38-54 (1979).

23. C.F. Curtiss and R.B. Bird, J. Chem. Phys, 74, 2016-2033 (1981); R.B. Bird, H.H. Saab, and C.F. Curtiss, J. Phys. Chem., 86, 1102-1106 (1982) and J. Chem. Phys., 77, 4747-4757 (1982); H.H Saab, R.B. Bird, and C.F. Curtiss, J. Chem. Phys., 77, 4758-4766 (1982).

24. C.F. Curtiss, R.B. Bird, and O. Hassager, Adv. Chem. Phys., 35, 31-117 (1976).

25. R.B. Bird and C.F. Curtiss, Physics Today, 37, 36-43 (January 1984).

26. J.G. Oldroyd, Proc. Roy. Soc., A200, 523-541 (1950).

LAGRANGIAN CONCEPTS FOR THE NUMERICAL ANALYSIS OF VISCOELASTIC FLOW

Bruce Caswell
Division of Engineering
Brown University
Providence, Rhode Island 02912
U.S.A.

1. Introduction

The last ten years has seen a steady increase in work on the numerical analysis of the flows of viscoelastic fluids. An excellent account of the status of the field as of 1983 can be found in the monograph by Crochet, Davies & Walters [1]. This surge of activity is driven by the ever increasing availability of computing power, and by the coincident development of constitutive equations which appear to describe qualitatively and sometimes quantitatively the known phenomena of viscoelasticity in fluids. This is not to say that a single equation has now gained acceptance by rheologists, and that work on constitutive theory will now subside. To be specific, for polymeric fluids a handful of constitutive equations are now regarded as appropriate for the analysis of flow problems. Indeed, one of the principal aims of the numerical work is to provide a means for comparing the predictions of these theories. One can regard computer simulation as an adjunct to the theoretical, analytical and experimental methods which have previously been brought to bear on the problem of polymer rheology.

2. Numerical Methods in Viscoelastic Flow

During the last few years the simulation of viscoelastic flow has been attempted by a number of workers [2,3,4,5,6,8,9] using a variety of methods based on both finite differences, finite elements, and boundary elements [10]. Until now the methods employed have been direct adaptations of those employed in the analysis of viscous fluids or elastic solids. The successes and failures of these efforts are well described in [1]; more recent work is to be found in the Journal of Non-Newtonian Fluid Mechanics. By and large success is associated with low values of the dimensionless relaxation time

(Deborah number, Weissenberg number), with convergence becoming more difficult as this parameter is increased. The explanation of this behaviour is not known at this time. It is known that with finite element and with finite difference methods success is most often achieved with systems characterized by self-adjoint operators. Fluid flow problems and their associated transport problems are notorious for the convergence barriers which appear when the governing equations become dominated by convective terms which are responsible for the loss of self-adjointness. Closely connected with these features is the possibility of change of type as the controlling parameter is varied. For the convection-diffusion equation the purely convective case reduces the system to first order hyperbolic. The results of recent investigations of Joseph [11] on the viscoelastic flows associated with several constitutive equations show some remarkable analogies to the transonic flow problem for the case when inertia is a significant parameter. For certain constitutive equations change of type may occur in the absence of inertia although these transitions have not been given a clear interpretation.

Numerical techniques for dealing with the convection problem have generally employed the "upwind" treatment of the convection term. This approach is essentially empirical, and has been controversial because of the inaccuracies it introduces. Recent attempts to provide a theoretical basis for the upwind technique can be found in the work of Hughes [12,13]. An alternative to upwinding is "symmetrization" in which the Galerkin statement of the problem is reformulated so that non-self-adjoint terms appear in conjunction with an adjustable parameter. Thus the degree of asymmetry becomes controllable. This approach has been applied to the convection-diffusion equation [14], but no systematic method of reformulation has been proposed. It is the opinion of the author that the viscoelastic flow problem with large elastic effects (high Deborah number) will require for its resolution the development of new methods which take into account the mathematical character of viscoelastic fluids. To this end a brief investigation of Lagrangian methods is presented below.

3. Lagrangian Concepts in Numerical Analysis

In fluid mechanics the Eulerian viewpoint is usually preferred since it best satisfies the requirements needed for the description of most flow problems. Furthermore experience shows that success in analysis is more likely when the governing equations are written in the Eulerian description. Indeed, the use of Lagrangian equations is known to be useful only in certain special cases. However, since the iterative solution procedures used in numerical analysis are based on consideration of small changes of field variables involving simple mathematics, it is worthwhile asking whether Lagrangian concepts can be of use in their development. The conservation laws, after all, are fundamentally Lagrangian.

Numerical approaches to fluid mechanics generally follow along paths previously trod by analytical investigators. This means that the equations to be cast into discrete form are invariably chosen to be the Eulerian equations which govern the continuous system. The accumulated experience of numerical simulation of the Navier-Stokes equation clearly demonstrates that performance is strongly affected by the rationale employed in the discretization process; this is expecially true with regard to numerical stability. For instance it has long been known that problems dominated by convection are numerically unstable when simulated by central finite difference schemes. The standard remedy usually prescribed to cure this condition is known as "upwinding" in which the V·grad( ) term is evaluated upstream from the point upon which the diffusion term is centered. This medicine is not free of side effects which appear in the form of reduced accuracy relative to what is attainable with central differences. Attempts to restore accuracy hinge upon willingness to incur the necessary overhead. In one dimension "upwind" has only two possible directions, in contrast with the infinite possibilities in two and three dimensions. In flows restricted to small departures from a single dominant direction low overhead, one dimensional, schemes may be made to perform with acceptable "crosswind diffusion" errors. The difficulties encountered with the upwinding errors stem in part from unwillingness to pay for the information necessary to properly determine from whence the "wind" is blowing. The only guaranteed way to find the upwind direction at a point is to ask where the corresponding material point resided a very short time ago. Thus it turns out that Lagrangian concepts must necessarily enter, and once the upstream position has been paid for the V·grad( ) terms disappear from view since the whole material derivative can then be approximated by simple differences. Schemes using finite differences along particle paths have been demonstrated for both the Navier-Stokes equation and the convection-diffusion equation in the numerical experiments of Bisshopp [15]. Finite element methods are more resistant to convective instabilities, but are eventually afflicted as the convective terms become stronger. When all elements are the same size and linear interpolation is used finite element methods can be shown to reduce to central finite difference schemes.

The aim of this paper is to study Lagrangian concepts which can be of use in the finite element simulation of viscoelastic flows. Viscoelastic materials are characterized by constitutive laws expressed as integro-differential equations written for a particle whose trajectory is followed in time. Here then Lagrangian concepts enter from the outset. Usually Lagrangian quantities are put into Eulerian form by well-known transformations, and these are added to the Eulerian conservation laws to complete the system. It is the purpose of this work to find out which Lagrangian quantities can be effectively retained without introducing the disadvantages of the material description.

4. Petrov-Galerkin Statements of Virtual Work

The following inner product notations will be utilized in the development of statements of virtual work:

$$<\mathbf{s},\mathbf{q}> = \int_V \mathbf{s}\cdot\mathbf{q}\ dV, \qquad <<\mathbf{s},\mathbf{q}>> = \int_{S_\ell} \mathbf{s}\cdot\mathbf{q}\ dA, \tag{1.1}$$

where $\mathbf{s}$ and $\mathbf{q}$ are any two tensors, and V is the spatial domain bounded by the surface S_ℓ. When the integration surface extends to the interior element boundaries the subscript i will be used. The process of construction of a statement of virtual work begins with a Galerkin integral statement over time and space,

$$\int_{-\infty}^{t} (<\rho(\mathbf{a}-\mathbf{b}),\mathbf{u}> + \int_V \mathrm{T}: \nabla\mathbf{u}\ dV - <<\boldsymbol{\tau},\mathbf{u}>>)dt' = 0. \tag{1.2}$$

Here T is the stress tensor, and u is the test function for velocity which vanishes on all parts of S where velocity is prescribed. On the remainder of S the traction $\boldsymbol{\tau}$ is imposed. In the first term ρ is the mass density, $\mathbf{a}$ the acceleration and $\mathbf{b}$ the body force. Except for the time integration equation (1.2) has the usual form of a Galerkin statement for spatially dependent fields (see Hughes [12]). The temporal evolution is usually treated by marching through time in steps Δt with spatial coordinates held constant which defines the Eulerian viewpoint.

Integration by parts of the stress term in (1.2) leads to

$$\int_{-\infty}^{t} \left\{<\mathrm{R},\mathbf{u}> + <<[\mathrm{T}\cdot\mathbf{n}],\mathbf{u}>>_i + <<\mathrm{T}\cdot\mathbf{n}-\boldsymbol{\tau}),\mathbf{u}>>_\ell\right\}dt' = 0 \tag{1.3}$$

where the momentum residual **R** is given by

$$\mathbf{R} = \rho(\mathbf{a}-\mathbf{b}) - \nabla\cdot\mathbf{T}, \qquad (1.4)$$

and $[\mathbf{T}\cdot\mathbf{n}]$ is the jump of the stress vector across element boundaries. Together with initial conditions appropriate to the flow this constitutes the weak statement of the problem.

For a fixed spatial domain the evolution of the system over $t - \Delta t$, t can be obtained approximately to a specified order of accuracy by subtraction of (1.3) at $t - \Delta t$ from its value at t with spatial positions held constant, and followed by division by Δt throughout. The "Euler-Lagrange" equations contained in (1.3) then reduce to

$$E(\Delta t)\langle\mathbf{R},\mathbf{u}\rangle = 0, \quad E(\Delta t)\langle\langle[\mathbf{T}\cdot\mathbf{n}],\mathbf{u}\rangle\rangle_i = 0, \quad E(\Delta t)\langle\langle(\mathbf{T}\cdot\mathbf{n}-\tau),\mathbf{u}\rangle\rangle_\ell = 0. \qquad (1.5)$$

In this paper the approximate Eulerian time operator is taken as

$$E(\Delta t) = 1 - (\Delta t/2)\partial/\partial t + 0(\Delta t^2). \qquad (1.6)$$

It is normal to take the unknown fields to be functions of time and space, but to regard the test functions **u** as dependent on the space variables alone. This is equivalent to the additional Petrov statements

$$\langle\mathbf{R},\partial\mathbf{u}/\partial t\rangle = 0, \quad \langle\langle[\mathbf{T}\cdot\mathbf{n}],\partial\mathbf{u}/\partial t\rangle\rangle_i = 0, \quad \langle\langle(\mathbf{T}\cdot\mathbf{n}-\tau), \partial\mathbf{u}/\partial t\rangle\rangle_\ell = 0. \qquad (1.7)$$

A statement which approximates (1.2) for one time step is obtained by reversing the by-parts integration of the stress terms in (1.5)

$$E(\Delta t)\left\{\langle\rho(\mathbf{a}-\mathbf{b}),\mathbf{u}\rangle + \int_V \mathbf{T}:\nabla\mathbf{u}\, dV - \langle\langle\tau,\mathbf{u}\rangle\rangle\right\} = 0(\Delta t^2). \qquad (1.8)$$

This is a standard working formula for the Eulerian time evolution of the unknown fields which in the incompressible case will be V,P the velocity and the pressure. It is conveniently represented by two-point central differences in time at a fixed spatial position. This standard Eulerian approach has been derived in a formal way for comparison with a new Lagrangian-Eulerian scheme derived below.

In a strictly Lagrangian analysis the time evolution follows a fixed portion of material. Since fluids may undergo infinite displacements this approach eventually fails because of severe element distortion as can be seen in the work of Hasseger [16]. The point of view taken here is basically Eulerian since the fields to be solved for are defined over a spatial domain. However, since the evolution of the system is traced with finite time steps it is possible to adopt a Lagrangian point of view towards the particles which occupy the domain during a single time step. Once the field variables have been obtained at the completion of the step a new set of particles is selected for the next step, and so on. By this means the distortion inherent in a totally Lagrangian description can be managed by adjustment of Δt. The material particles are selected to be those which coincide exactly with the spatial system or control volume at time t. At an earlier time t - Δt the material system will lie partly outside the spatial domain, as depicted in Figure 1.

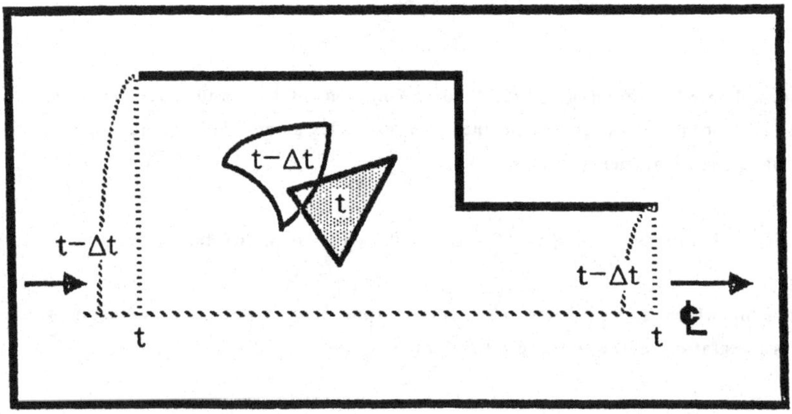

Figure 1.

Since the particles occupy the spatial domain at time t, the time integration of (1.2) and (1.3) can be thought of as being carried out either from the Eulerian viewpoint, already presented above, or from the Lagrangian viewpoint. The Lagrangian integration is effected by transformation of the volume and surface integrals to a fixed material reference, followed by approximate time integration with fixed material coordinates, and then followed by transformation back to the spatial system. The resulting "Euler-Lagrange" equations can be written in the same form as (1.5) with the Eulerian time operator replaced by the Lagrangian time operator

$$L(\Delta t) = 1 - (\Delta t/2)\, D\ /Dt + O(\Delta t^2), \tag{1.9}$$

where D /Dt stands for the material derivative. As with the Eulerian case a time marching scheme requires an additional assumption for the time dependence of the test functions **u**. The natural choice appears to be (1.7) with $\partial \mathbf{u}/\partial t$ replaced by $D\mathbf{u}/Dt$,

$$\langle \mathbf{R}, D\mathbf{u}/Dt \rangle = 0, \quad \langle\langle [\mathbf{T}\cdot\mathbf{n}], D\mathbf{u}/Dt \rangle\rangle_i = 0, \quad \langle\langle (\mathbf{T}\cdot\mathbf{n}-\boldsymbol{\tau}), D\mathbf{u}/Dt \rangle\rangle_\ell = 0. \tag{1.10}$$

In effect the test functions are taken to be materially constant. With the use of (1.10) the Lagrangian counterpart of the "Euler-Lagrange" equations (1.5) is found to be

$$\langle L(\Delta t)\mathbf{R}, \mathbf{u} \rangle = 0, \quad \langle\langle [\overset{*}{\mathbf{T}}\cdot\mathbf{n}], \mathbf{u} \rangle\rangle_i = 0, \quad \langle\langle (\overset{*}{\mathbf{T}}\cdot\mathbf{n}-\overset{*}{\boldsymbol{\tau}}), \mathbf{u} \rangle\rangle_\ell = 0, \tag{1.11}$$

where

$$\overset{*}{\mathbf{T}} = \mathbf{T} - (\Delta t/2)(\overset{\triangledown}{\mathbf{T}} + (\nabla\mathbf{V})^+\cdot\mathbf{T}), \tag{1.12}$$

the transpose is denoted by $(\ )^+$, and $\triangledown$ is the upper convected derivative such that

$$\overset{\triangledown}{\mathbf{T}} = D\mathbf{T}/Dt - (\nabla\mathbf{V})^+\cdot\mathbf{T} - \mathbf{T}\cdot\nabla\mathbf{V} + \nabla\cdot\mathbf{V}\mathbf{T}. \tag{1.13}$$

The first of the "Euler-Lagrange" equations (1.11) is a Galerkin statement of impulse. Its energy equivalent which approximates (1.2) for one time step is obtained from (1.11) by following the analogous steps used to arrive at (1.8) from (1.5),

$$\left\{ \langle L(\Delta t)(\rho(a-b)),u\rangle + \langle \overset{*}{T},\nabla u\rangle - \langle\langle \overset{*}{\tau},u\rangle\rangle_{\ell} \right\} = 0(\Delta t^2). \tag{1.14}$$

The stress flux which appears in (1.14) is the Eulerian form of the material derivative of the Piola stress, i.e. its components transform according to

$$\overline{JX^K_{,k}T^{ik}} = JX^K_{,k}(\overset{\nabla}{T}{}^{ik} + V^i_{,\ell}T^{\ell k}), \tag{1.15}$$

where the superimposed dot indicates the material derivative. Here X^K is the usual notation for components of material coordinates, and J is the volume ratio.

$$L(\Delta t)\mathbf{a} = \mathbf{a}(\mathbf{X},t_0) = \Delta\mathbf{V}/\Delta t, \quad t_0 = t - \Delta t/2. \tag{1.16}$$

The velocities of the particle **X** must be interpolated from Eulerian fields determined over the whole space at every time step:

$$\Delta\mathbf{V} = \mathbf{V}(\mathbf{X},t) - \mathbf{V}(\mathbf{X},t-\Delta t) = \mathbf{V}(\mathbf{x},t) - \mathbf{V}(\mathbf{x}-\Delta\mathbf{x},t-\Delta t), \tag{1.17}$$

where $\mathbf{x}$ and $\mathbf{x} - \Delta\mathbf{x}$ are the positions of **X** at t and $t - \Delta t$ respectively. Notice that the difference formula in (1.16) is exact; the unavoidable error enters in (1.16) in the estimation of the particle displacement. In the scheme outlined by Caswell [17] velocities at times $t - \Delta t$, and $t - 2\Delta t$ are stored for this purpose in order to be able to obtain Δx to $0(\Delta t^2)$.

With the help of (1.15) the stress term $\overset{*}{T} : \nabla\mathbf{u}$ has the interpretation

$$\overset{*}{T} : \nabla \mathbf{u} = J(\mathbf{z},\mathbf{X})(\partial X^K/\partial z^k)T^{i\ell}(\mathbf{X},t_0)u_{i,K} + 0(\Delta t^2) \tag{1.18}$$

$$= \left\{ J(\mathbf{y},\mathbf{X})(\partial X^K/\partial y^\ell)T^{i\ell}(\mathbf{X},t-\Delta t) + (\Delta t/2)\ \overline{J X^K_k T^{i\ell}}^{\,\cdot} \right\} u_{i,K} + 0(\Delta t^2),$$

$$= (J(\mathbf{y},\mathbf{x})(\partial x^k/\partial y^\ell)T^{i\ell}(\mathbf{X},t-\Delta t) + T^{ik}(\mathbf{X},t))/2)u_{i,k} + 0(\Delta t^2).$$

In these equations

$$\mathbf{y} = \mathbf{x}(\mathbf{X},t-\Delta t) = \mathbf{x} - \Delta\mathbf{x}, \quad \mathbf{z} = \mathbf{x}(\mathbf{X},t_0). \tag{1.19}$$

The use of (1.14) as the basis for an algorithm for the simulation of viscoelastic flows requires the simultaneous solution of a constitutive equation for the stress. The constraint of constant volume must also be included, but since this aspect is well-known it will be omitted in this paper for the sake of brevity. For consistency the isotropic pressure is not represented explicitly in any of these equations even though the pressure is considered to be an unknown.

5. Algorithm for Viscoelastic Flow

The a lgorithm for viscoelastic flow will be illustrated for the case of the upper convected Maxwell fluid with viscosity η and relaxation time λ expressed in differential form as

$$\overset{\nabla}{T} = (\eta/\lambda)A - (1/\lambda)T, \quad A = \nabla V + (\nabla V)^+. \tag{2.1}$$

It is assumed that the stress field has been determined at t - Δt. Hence the first term of (1.18b) can be obtained by interpolation once the particle displacement has been estimated. The stress rate in the second term of (1.18b) is eliminated with the constitutive equation (2.1), and after some manipulation it can be shown that the stress term in (1.14) becomes

$$\overset{*}{T}: \nabla u = \frac{\partial x^{\ell}}{\partial y^{k}} \underset{-}{T}^{ik} \frac{(\delta_{i}^{m} + \Delta t V_{,i}^{m})}{(1 + \Delta t/2\lambda)} u_{m,\ell} + \frac{(\eta t/2\lambda)}{2(1 + \Delta t/2\lambda)} \frac{\partial x^{\ell}}{\partial y^{k}} (\underset{-}{A}^{ik} + A^{i\ell}) u_{i,\ell}. \quad (2.2)$$

In this equation quantities with the - subscript are evaluated at y, t - Δt, and therefore are known. The unsubscripted quantities are then expressible in terms of the unknown velocity at **x**,t. As a basis for an algorithm for the flow of a Maxwell fluid (2.2) is almost the same as that proposed by Caswell [17]. The development given here is more systematic, and makes clear that a time marching scheme always contains an assumption on the time dependence of the test functions. The materially constant test functions of this work lead to the replacement of the usual Galerkin-momentum statement with a Galerkin-impulse statement. The algorithm for the calculation of transient flows which makes use of (2.2) consists of the following:

i) Storage is provided for velocity fields at two prior times, and for the stress field at the previous time. At a field point x the displacement Δx at t - Δt of the corresponding particle is evaluated with the use of the two velocity fields. The stress and any other field variables associated with the particle at t - Δt are interpolated at x - Δx.

ii) The information generated in (i) is sufficient to set up (2.2) which becomes a term in the overall impulse statement (1.14). Inspection of these equations shows that the only unknowns are the velocity and pressure at the current time t.

iii) Once V,P at t have been determined an expression very similar to (2.2) is used to update the stress field to its value at t. In this step the particle displacements are determined from the new velocity to give an improved estimate. Together (ii) and (iii) have the character of predictor-corrector schemes, and indeed it may be necessary to repeat these steps within a time step to achieve the full accuracy of the scheme.

Numerical experiments are currently being carried out with this Eulerian-Lagrangian algorithm. The difficulties encountered at this stage appear to be related to the rule used to interpolate the pressure field. The rules employed in standard finite element formulations of the Stokes equation no longer work since in this formulation the viscous terms become nonlinear.

In the introduction two approaches, upwinding and symmetrization, for dealing with systems governed by highly nonsymmetric operators were outlined. Clearly, the integration along particle paths is an exact implementation of the notion of upwinding, with the consequence that the inertia term in (1.14) is completely symmetrized. For the Navier-Stokes fluid the viscous terms become nonlinear, and when linearized nonsymmetric. However, the asymmetry appears with Δt, and is thereby controllable. For the Maxwell fluid the stress term (2.2) also has a controllable asymmetry. Thus it appears that with regard to asymmetry the Eulerian-Lagrangian approach lies between the highly nonsymmetric Eulerian formulations and the completely symmetric Lagrangian formulation of Hasseger [16] with the displacement field as the unknown.

Acknowledgement

Support for this work by the National Science Foundation under Grant MEA-8217200 is gratefully acknowledged.

References

1. M. J. Crochet, A. R. Davies, K. Walters, "Numerical simulation of non-Newtonian flow", Rheology series, 1, Elsevier, Amsterdam, 1984.

2. R. Keunings and M. J. Crochet, "Numerical simulation of the flow of a viscoelastic fluid through an abrupt contraction", J. Non-Newtonian Fluid Mech., **14**, 279 (1984).

3. B. Caswell and M. Viriyayuthakorn, "Finite element simulation of die swell for a Maxwell fluid", J. Non-Newtonian Fluid Mech. **12**, 13 (1983).

4. P. W. Yeh, M. E. Kim-E, R. C. Armstrong and R. A. Brown, "Multiple solutions in the calculation of axisymmetric contraction flow of an upper convected Maxwell fluid", J. Non-Newtonian Fluid Mech., **16**, 173 (1984).

5. S. L. Josse and Bd. A. Finlayson, "Reflections on the numerical viscoelastic flow problem", J. Non-Newtonian Fluid Mech., **16**, 13 (1984).

6. D. S. Malkus and B. Bernstein, "Flow of a Curtiss-Bird fluid over a transverse slot using the finite element drift function method", J. Non-Newtonian Fluid Mech., **77** (1984).

7. A. R. Davies, "Numerical filtering and the high Weissenberg number problem", J. Non-Newtonian Fluid Mech., **16**, 195 (1984).

8. R. Keunings, "An algorithm for the simulation of transient viscoelastic flows with free surfaces", J. Comput. Phys., 1985 (in press).

9. R. Keunings, "Mesh refinement analysis of the flow of a Maxwell fluid through an abrupt contraction", 4th Int. Conf. on Numerical Methods in Laminar and Turbulent Flow, Swansea, July 1985.

10. M. B. Bush, J. F. Milthorpe and R. I. Tanner, "Finite element and boundary element methods for extrusion computations", J. Non-Newtonian Fluid Mech., **16**, 37 (1984).

11. D. D. Joseph, "Hyperbolic phenomena in the flow of viscoelastic fluids", presented at the Symposium on Viscoelasticity and Rheology, Madison, Wis., October 1984.

12. A. N. Brooks and T. J. R. Hughes, "Streamline upwind/Petrov-Galerkin formulations for convection dominated flows with particular emphasis on the incompressible Navier-Stokes equations", Comp. Meth. Appl. Mech. Eng., **32**, 97 (1984).

13. T. J. R. Hughes and T. E. Tezduyar, "Finite element methods for first-order hyperbolic sysems with particular emphasis on the compressible Euler equations", Comp. Meth. Appl. Mech. Eng., **45**, 217 (1984).

14. J. W. Barrett, K. W. Morton, "Approximate symmetrization and Petrov-Galerkin methods for diffusion-convection problems", Comp. Meth. Appl. Mech. Eng., **45**, 97 (1984).

15. F. C. Bisshopp, Private Communication.

16. O. Hasseger and C. Bisgaard, "A Lagrangian finite element method for the flow of non-Newtonian liquids", J. Non-Newtonian Fluid Mech., **12**, 153 (1983).

17. B. Caswell, "An Eulerian-Lagrangian formulation for the numerical analysis of viscoelastic flow", Advances in Rheology, 1. Theory, ed. by B. Mena, A. Garcia-Rejon and C. Rangel-Nafaile, Universidad Nacional Autonoma de Mexico, Mexico, 259 (1984).

Solutions with Shocks for Conservation Laws with Memory

C. M. Dafermos

Lefschetz Center for Dynamical Systems
Division of Applied Mathematics
Brown University
Providence, Rhode Island 02912

0. Motivation

The equations of motion of a one-dimensional body with unit reference density and zero body force, in Lagrangian coordinates, read

$$\begin{cases} \partial_t u(x,t) - \partial_x v(x,t) = 0 \\ \partial_t v(x,t) - \partial_x \sigma(x,t) = 0 \end{cases} \tag{0.1}$$

where u is deformation gradient, v is velocity, and σ denotes stress.

When the body is elastic, the stress at the material point x and time t is determined solely by the value of deformation gradient at (x,t) via a constitutive relation

$$\sigma(x,t) = f(u(x,t)) \;. \tag{0.2}$$

Under the standard assumption $f'(u) > 0$, (0.1), (0.2) yield a strictly hyperbolic system for which the Cauchy problem has been studied extensively: when the initial data are smooth, a classical solution starts out at $t = 0$ but eventually breaks down in a finite time, with the formation of shocks (cf. [14, 13]). When the initial data have small total variation, an admissible weak solution exists, globally in time, in the class BV of functions of bounded variation (cf. [9]). Furthermore, if the system is *genuinely nonlinear*, in the sense that $f''(u) \neq 0$ on its domain of definition, then a globally defined admissible BV solution exists under initial data that are merely bounded measurable and have small oscillation (cf. [10]). There is strong evidence that globally defined weak solutions exist in the class of bounded measurable functions, under any bounded measurable initial data, though this has been strictly established only when $(u - \bar{u})f''(u) \geq 0$ for some $\bar{u}$ and all u (cf. [7, 20]).

It is of interest to compare and contrast the behavior of elastic bodies with the behavior of materials with *fading memory* (cf. [22]), in which the stress $\sigma(x,t)$ at the material point x and time t is determined by the entire history $u^{(t)}(x,\cdot)$ of deformation gradient at x, defined by $u^{(t)}(x,\tau) = u(x,t-\tau)$, $0 \leq \tau < \infty$, through a functional with appropriate smoothness properties.

For concreteness, let us consider the simple model

$$\sigma(x,t) = f(u(x,t)) - z(x,t) \tag{0.3}$$

where

$$z(x,t) = \int_{-\infty}^{t} K(t-\tau)\, g(u(x,\tau))\, d\tau . \tag{0.4}$$

The study of acceleration wave propagation in media of this type (cf. [1]) suggests that the memory exerts a rather weak damping influence. This is corroborated by the observation (cf. [15]) that when $g(u) \equiv f(u)$ the system (0.1), (0.3), (0.4) may be rewritten in the form

$$\begin{cases} \partial_t u(x,t) - \partial_x v(x,t) = 0 \\ \partial_t v(x,t) - \partial_x f(u(x,t)) + K(0)v(x,t) - \int_{-\infty}^{t} L'(t-\tau)\, v(x,\tau)\, d\tau = 0 , \end{cases} \tag{0.5}$$

where $L(t)$ denotes the resolvent kernel of $K(t)$. It is clear that in this case memory acts just like frictional damping.

The Cauchy problem for (0.1), (0.3), (0.4) has been studied by several authors: Assuming $K(t)$ satisfies appropriate assumptions, it has been shown (cf. [6, 12]) that when the initial data are smooth and "small" dissipation prevails over the destabilizing action of nonlinear instantaneous elastic response and, as a result, a smooth solution exists globally in time. On the other hand, when $f(u)$ is nonlinear and the initial data are "large" then the destabilizing action of nonlinearity prevails and solutions break down in a finite time (cf. [4,18]).

One should expect that the theory of weak solutions for (0.1), (0.3), (0.4) parallels the theory of weak solutions for (0.1), (0.2). This, however, has not been verified at the present time. The reason is that (0.1), (0.3), (0.4) possesses neither self-similar solutions, of the type used as building blocks in the construction of BV solutions for (0.1), (0.2) by the *random choice* method, nor entropies, like those that play a crucial role in establishing the existence of bounded measurable solutions for (0.1), (0.2) by the method of *compensated compactness*.

As is clear from (0.5), in the special case $f(u) \equiv g(u)$ $\partial_x z(x,t)$ turns out to be smoother than $\partial_x f(u(x,t))$, i.e., integration with respect to t offsets differentiation with respect to x. Establishing this property *a priori* in the general case would be a major step towards settling the issue of weak solutions for (0.1), (0.3), (0.4).

A different perspective in the difficulty of the problem is provided by the observation that when $K(t) = \exp(-\alpha t)$ then (0.1), (0.3), (0.4) is equivalent to the system

$$(0.6) \qquad \begin{cases} \partial_t u(x,t) - \partial_x v(x,t) = 0 \\ \partial_t v(x,t) - \partial_x f(u(x,t)) - \partial_x z(x,t) = 0 \\ \partial_t z(x,t) = g(u(x,t)) - \alpha z(x,t) . \end{cases}$$

Under the hypothesis $f'(u) > 0$, (0.6) is strictly hyperbolic but it is nonhomogeneous and has one linearly degenerate characteristic field.

In order to prepare the ground for a future attack on this difficult problem, we shall examine here a simpler model which, however, exhibits the same features as (0.1), (0.3), (0.4). We will study weak solutions for a conservation law with memory which bears to (0.1), (0.3), (0.4) the same relationship that the single conservation law bears to the hyperbolic system (0.1), (0.2).

Acknowledgement: This research was done while I was visiting the Institute of Mathematics and its Applications at the University of Minnesota. I am indebted to my colleagues there and, in particular, to Jerald Ericksen, David Kinderlehrer, George Sell, and Hans Weinberger for giving me the opportunity to work in a pleasant and stimulating environment. The work was also partially supported by the National Science Foundation under grant no. DMS-8205355 and by the U. S. Army under contract no. DAAG-29-83-K-0029.

1. Introduction

We consider the initial-value problem

(1.1) $$\partial_t u(x,t) + \partial_x \sigma(x,t) = 0\ , \qquad -\infty < x < \infty\ , \quad 0 < t < \infty\ ,$$

(1.2) $$u(x,0) = u_0(x)\ , \qquad -\infty < x < \infty\ ,$$

where $\sigma(x,t)$ is determined by the "history" of $u(x,\cdot)$ on $[0,t]$ via the functional

(1.3) $$\sigma(x,t) = f(u(x,t)) - z(x,t)\ ,$$

(1.4) $$z(x,t) = \int_0^t K(t-\tau)\, g(u(x,\tau))\, d\tau\ .$$

The objective of our investigation is to compare and contrast the behavior of solutions of (1.1), (1.3), (1.4) with the behavior of solutions of the hyperbolic conservation law

$$\partial_t u(x,t) + \partial_x f(u(x,t)) = 0\,, \qquad -\infty < x < \infty\,, \quad 0 < t < \infty\,. \tag{1.5}$$

As is well known, when $f(u)$ is nonlinear, smooth solutions of (1.5), (1.2) generally break down in a finite time, with the formation of shocks. The theory of weak solutions is well understood (cf. [19]): when $u_0(x)$ is bounded measurable there exists a unique admissible solution $u(x,t)$ of (1.5), (1.2) on $(-\infty,\infty) \times [0,\infty)$ in the class of bounded measurable functions. Furthermore, if $u_0(x)$ has bounded variation then $u(x,t)$ is a function of class BV. Perhaps the most striking property is that when $f''(u)$ does not vanish on $(-\infty,\infty)$ then $u(x,t)$ has locally bounded variation, even when $u_0(x)$ is merely bounded. In that situation, $TV_x u(x,t)$ may blow up to infinity at the rate t^{-1}, as $t \downarrow 0$.

When $g(u)$ and the kernel $K(t)$ satisfy appropriate conditions, the effect of memory is dissipative and, as a result, smooth solutions of (1.1), (1.3), (1.4), (1.2) exist globally in time, providing $u_0(x)$ is smooth and "small" (cf. [17, 3, 21]). Unfortunately, the damping induced by fading memory is weak and so smooth solutions break down in finite time when $u_0(x)$ is "large" (cf. [16, 3, 21]).

Existence for globally defined admissible weak solutions for (1.1), (1.3), (1.4), (1.2) in the class of locally bounded measurable functions, under bounded measurable initial data, has been established in [5], by the method of compensated compactness. Existence of BV solutions, under initial data with bounded variation, is shown in [11], albeit only in the special case $f(u) \equiv g(u)$.

The intent here is to look into the problem of existence of BV solutions, in the general case $f(u) \not\equiv g(u)$, under initial data that have bounded variation or are merely bounded. Actually, here we will only establish appropriate *a priori* estimates; presumably, the construction of solutions may then be effected by various techniques, e.g., via difference approximations. The assumptions and results are recorded in Section 2 and the proofs, by the method of generalized characteristics (cf. [2]), are given in subsequent sections.

2. Statement of Results

We consider the initial-value problem (1.1), (1.3), (1.4), (1.2) under the following assumptions: $f(u)$ is a C^2 real-valued function on $(-\infty,\infty)$, normalized by $f(0) = 0$, which is strictly increasing,

(2.1) $$f'(u) > 0 \,, \qquad -\infty < u < \infty \,,$$

surjective,

(2.2) $$f(u) \longrightarrow \pm\infty \,, \qquad \text{as} \quad u \longrightarrow \pm\infty \,,$$

and strictly convex,

(2.3) $$f''(u) > 0 \,, \qquad -\infty < u < \infty \,;$$

$g(u)$ is a C^1 real-valued function on $(-\infty,\infty)$, similarly normalized by $g(0) = 0$, which grows at infinity no faster than $f(u)$, i.e.,

(2.4) $$|g(u)| \leqslant |f(u)| \,, \qquad -\infty < u < \infty \,.$$

The kernel $K(t)$ is C^2 on $[0,\infty)$. The initial data $u_0(x)$ are left-continuous, periodic,

(2.5) $$u_0(x+1) = u_0(x) \quad , \qquad -\infty < x < \infty \,,$$

and have bounded variation over $[0,1]$.

Assumption (2.4) ensures that the "memory" response does not dominate the "instantaneous elastic" response while (2.1) guarantees that hyperbolic waves propagate at non-zero speed and hence do not resonate with the memory term. The condition (2.3) of "genuine nonlinearity" induces spreading of rarefaction waves thus generating solutions with (locally) bounded variation out of initial data of (possibly) infinite variation. The assumption (2.5) of periodicity is only made for convenience; more general initial data may be treated by the same method.

In the sequel, $u(x,t)$ will denote a solution of (1.1), (1.3), (1.4), (1.2) which satisfies the following assumptions: for each fixed $t \in (0,\infty)$, $u(\cdot,t)$ is a periodic,

(2.6) $$u(x+1,t) = u(x,t), \qquad -\infty < x < \infty, \quad 0 < t < \infty,$$

left-continuous function of locally bounded variation such that

(2.7) $$u(x+,t) \leq u(x,t), \qquad -\infty < x < \infty, \quad 0 < t < \infty.$$

For each fixed $x \in (-\infty,\infty)$, the function $u(x,\cdot)$ has locally bounded variation on $[0,\infty)$. Furthermore, the function that carries t into $u(\cdot,t)$ is (locally) Lipchitz continuous in $L^1_{loc}(-\infty,\infty)$ on $[0,\infty)$ and the function that carries x into $u(x,\cdot)$ is Lipschitz continuous in $L^1_{loc}[0,\infty)$ on $(-\infty,\infty)$.

Assumption (2.7) expresses the familiar shock admissibility condition (sometimes called "entropy" condition) which is well motivated in the context of the theory of the hyperbolic conservation law (1.5) (cf. [19]).

By virtue of (1.4), the above assumptions imply, in particular, that $z(x,t)$ is (locally) Lipschitz continuous on $(-\infty,\infty) \times [0,\infty)$.

Our goal here is to establish *a priori* bounds that justify the smoothness conditions on $u(x,t)$ imposed above. We fix any $T > 0$ and consider the restriction of $u(x,t)$ on $(-\infty,\infty) \times [0,T]$.

Proposition 2.1. *There is* M *which depends solely on* T *and on* $\sup|u_0(\cdot)|$ *such that*

(2.8) $$|u(x,t)| \leq M, \qquad -\infty < x < \infty, \quad 0 \leq t \leq T.$$

Proposition 2.2. *There is* N *which depends solely on* T , *on* $\sup|u_0(\cdot)|$ *and on the total variation of* $u_0(x)$ *over* [0,1] *such that*

$$\underset{[0,1]}{TV}\, u(\cdot,t) \leq N\,, \qquad 0 \leq t \leq T\,, \tag{2.9}$$

$$\underset{[0,T]}{TV}\, u(x,\cdot) \leq N\,, \qquad -\infty < x < \infty\,, \tag{2.10}$$

$$\int_0^1 |u(x,t) - u(x,s)|\,dx \leq N(t-s)\,, \qquad 0 \leq s < t \leq T\,, \tag{2.11}$$

$$\int_0^T |u(y,t) - u(x,t)|\,dt \leq N(y-x)\,, \qquad -\infty < x < y < \infty\,. \tag{2.12}$$

The main result of the paper is contained in the following proposition, which states that when the "memory" response is appropriately dissipative then the total variation of the solution, away from $t = 0$, is bounded independently of the variation of the initial data.

Proposition 2.3. *In addition to* (2.1), (2.2), (2.3), (2.4), *assume*

$$\frac{g'(u)}{f'(u)} \text{ is monotone on } (-\infty,\infty)\,, \tag{2.13}$$

$$K(t) \geq 0\,, \qquad 0 \leq t < \infty\,, \tag{2.14}$$

$$K'(t) \leq 0\,, \qquad 0 \leq t < \infty\,. \tag{2.15}$$

Then there is M *which depends solely on* T *and on* $\sup|u_0(\cdot)|$ *such that*

$$\underset{[0,1]}{TV}\, u(\cdot,t) \leq \frac{M}{t}\,, \qquad 0 < t \leq T\,. \tag{2.16}$$

It is clear that (2.16) induces compactness that would allow us to pass, via completion, from solutions with initial data of bounded variation to solutions with initial data that are merely bounded.

The proof of Propositions 2.1, 2.2, and 2.3, given in subsequent sections, will be based on the theory of generalized characteristics for (1.1), (1.3), (1.4) which is outlined in Section 3.

3. Generalized Characteristics

In this section we explain how characteristics for (1.1), (1.3), (1.4) can be realized in the context of weak solutions, within the function class considered here. For a thorough discussion of related ideas, the reader is referred to [2].

Definition 3.1. A *characteristic*, associated with the solution u, emanating from a point (x,t) in $(-\infty,\infty) \times (0,T]$ is a Lipschitz function $\xi(\cdot)$ defined on $[0,t]$ such that $\xi(t) = x$ and

$$\dot{\xi}(\tau) \in \left[f'\left(u(\xi(\tau)+,\tau)\right),\ f'\left(u(\xi(\tau)-,\tau)\right)\right], \quad \text{a.e. on } [0,t] \ . \tag{3.1}$$

By the general theory of differential inclusions (cf. [8]), from any point (x,t) in $(-\infty,\infty) \times (0,T]$ there emanates a funnel of characteristics confined between a *minimal* and a *maximal* one. The special properties of these extremal characteristics will play a central role in the analysis here.

Lemma 3.1. *Let* $\xi(\cdot)$ *denote the minimal characteristic emanating from a point* (x,t) *in* $(-\infty,\infty) \times (0,T]$. *Then*

$$\dot{\xi}(\tau) = f'\left(u(\xi(\tau)-,\tau)\right) \qquad \text{a.e. on } [0,t] \ , \tag{3.2}$$

$$u(\xi(\tau)+,\tau) = u(\xi(\tau)-,t) \qquad \text{a.e. on } [0,t] \ . \tag{3.3}$$

Proof. The interpretation of (3.2) is clear: being minimal, $\xi(\cdot)$ selects the maximum allowable speed of propagation which, by virtue of (3.1), is $f'\left(u(\xi(\tau)-,\tau)\right)$. For a proof, see [2].

Now fix $0 \leq t_1 < t_2 \leq t$ and apply the measure on the left-hand side of (1.1) to the arc $\{(\xi(\tau),\tau) : t_1 \leq \tau \leq t_2\}$. This yields

$$\int_{t_1}^{t_2} \Big\{ f\big[u(\xi(\tau)+,\tau)\big] - f\big[u(\xi(\tau),\tau)\big] - \dot{\xi}(\tau)[u(\xi(\tau)+,\tau) - u(\xi(\tau),\tau)]\Big\} d\tau = 0 . \tag{3.4}$$

It follows that the integrand in (3.4) vanishes almost everywhere on $[0,t]$. Combining this with (3.2) and recalling (2.3), we deduce (3.3). This completes the proof. ∎

On $(-\infty,\infty) \times [0,T]$ we define the functions

$$p(x,t) := \int_0^t K'(t-\tau) g(u(x,\tau)) d\tau , \tag{3.5}$$

$$h(x,t) := K(0) g(u(x,t)) + p(x,t) . \tag{3.6}$$

We note that $p(x,t)$ is Lipschitz on $(-\infty,\infty) \times [0,T]$ and, for each fixed $x \in (-\infty,\infty)$, $\partial_t z(x,t) = h(x,t)$ for almost all $t \in [0,T]$.

Lemma 3.2. *Let* $\xi(\cdot)$ *denote the minimal characteristic emanating from a point* (x,t) *in* $(-\infty,\infty) \times (0,T]$. *Then*

$$\sigma(\xi(s),s) = \sigma(x,t) + \int_s^t h(\xi(\tau),\tau) d\tau , \qquad 0 < s \leq t , \tag{3.7}$$

$$f\big[u_0(\xi(0))\big] \leq \lim_{s \downarrow 0} \sigma(\xi(s),s) \leq f\big[u_0(\xi(0)+)\big] . \tag{3.8}$$

Proof. We fix $\epsilon > 0$. On $[0,t]$ we define the functions

$$u_\epsilon(\tau) := \frac{1}{\epsilon} \int_{\xi(\tau)-\epsilon}^{\xi(\tau)} u(x,\tau) dx , \tag{3.9}$$

$$z_\epsilon(\tau) := \frac{1}{\epsilon} \int_{\xi(\tau)-\epsilon}^{\xi(\tau)} z(x,\tau) dx , \tag{3.10}$$

(3.11) $$\sigma_\epsilon(\tau) := f(u_\epsilon(\tau)) - z_\epsilon(\tau) ,$$

(3.12) $$h_\epsilon(\tau) := \frac{1}{\epsilon} \int_{\xi(\tau)-\epsilon}^{\xi(\tau)} h(x,\tau)\,dx ,$$

which are Lipschitz continuous, by account of our assumptions on u (cf. Section 2), and tend, as $\epsilon \downarrow 0$, to $u(\xi(\tau),\tau)$, $z(\xi(\tau),\tau)$, $\sigma(\xi(\tau),\tau)$ and $h(\xi(\tau),\tau)$, respectively, for all $\tau \in [0,t]$.

We fix any $\tau \in (0,t)$ at which $u_\epsilon(\cdot)$, $z_\epsilon(\cdot)$, $\sigma_\epsilon(\cdot)$, $\xi_\epsilon(\cdot)$ are differentiable and, at the same time, it is

(3.13) $$u(\xi(\tau)+,\tau) = u(\xi(\tau),\tau) ,$$

(3.14) $$\dot\xi(\tau) = f'\left[u(\xi(\tau),\tau)\right] .$$

Then, recalling (3.10), (1.4), (3.6), (3.5), and (3.12), we obtain

(3.15) $$\dot z_\epsilon(\tau) = \frac{1}{\epsilon}\left\{z(\xi(\tau),\tau) - z(\xi(\tau)-\epsilon,\tau)\right\}\dot\xi(\tau) + h_\epsilon(\tau) .$$

Furthermore, using (3.9), (1.1), (1.3), (1.4), (3.14), and (2.3) we get

(3.16) $$\begin{aligned}\dot u_\epsilon(\tau) = {} & \frac{1}{\epsilon}\left\{z(\xi(\tau),\tau) - z(\xi(\tau)-\epsilon,\tau)\right\} \\ & - \frac{1}{\epsilon}\Big\{f\left[u(\xi(\tau),\tau)\right] - f\left[u(\xi(\tau)-\epsilon,\tau)\right] \\ & \qquad - f'\left[u(\xi(\tau),\tau)\right]\left[u(\xi(\tau),\tau) - u(\xi(\tau)-\epsilon,\tau)\right]\Big\} \\ \geq {} & \frac{1}{\epsilon}\left\{z(\xi(\tau),\tau) - z(\xi(\tau)-\epsilon,\tau)\right\} .\end{aligned}$$

Combining (3.16) with (3.11) and (3.15) yields

(3.17) $$\dot\sigma_\epsilon(\tau) \geq \frac{1}{\epsilon}\left\{z(\xi(\tau),\tau) - z(\xi(\tau)-\epsilon,\tau)\right\}\left\{f'(u_\epsilon(\tau)) - \dot\xi(\tau)\right\} - h_\epsilon(\tau) .$$

For $s \in [0,t)$, we integrate (3.17) over (s,t) and then let $\epsilon \downarrow 0$. Recalling that z is Lipschitz and using (3.2), we deduce

$$(3.18)\qquad \sigma(x,t) - \sigma(\xi(s),s) \geq -\int_s^t h(\xi(\tau),\tau)\,d\tau\,, \qquad 0 \leq s \leq t\,.$$

Next we fix $\epsilon < 0$, redefine $u_\epsilon(\cdot)$, $z_\epsilon(\cdot)$, $\sigma_\epsilon(\cdot)$, $h_\epsilon(\cdot)$ on $[0,t]$ through (3.9), (3.10), (3.11), (3.12) and retrace the steps of the analysis described above thus arriving, in the place of (3.18), at the inequality

$$(3.19)\qquad \sigma(x+,t) - \sigma(\xi(s)+,s) \leq -\int_s^t h(\xi(\tau)+,\tau)\,d\tau\,, \qquad 0 \leq s \leq t\,.$$

We note that the right-hand sides of (3.18), (3.19) are equal, by account of (3.6), (3.5), and (3.3).

Assume first

$$(3.20)\qquad u(x+,t) = u(x,t)\,,$$

in which case $\sigma(x+,t) = \sigma(x,t)$. By (1.3), (2.1), and (2.7) it is $\sigma(\xi(s)+,s) \leq \sigma(\xi(s),s)$, $0 < s < t$, and so both (3.18), (3.19) must be satisfied as equalities on $(0,t]$, i.e., (3.7) holds. Furthermore, applying (3.18), (3.19) for $s = 0$ and letting $s \downarrow 0$ in (3.7) we deduce (3.8).

Assume now

$$(3.21)\qquad u(x+,t) < u(x,t)\,.$$

We consider an increasing sequence $\{x_n\}$ such that $u(x_n+,t) = u(x_n,t)$, $n = 1,2,\cdots$, and $x_n \uparrow x$ as $n \to \infty$. If $\xi_n(\cdot)$ denotes the minimal characteristic emanating from (x_n,t) then, as shown above,

$$(3.22)\qquad \sigma(\xi_n(s),s) = \sigma(x_n,t) + \int_s^t h(\xi_n(\tau),\tau)\,d\tau\,, \qquad 0 < s \leq t\,.$$

At the same time, by minimality, $\xi_1(\tau) \leq \xi_2(\tau) \leq \cdots \leq \xi(\tau)$, $0 \leq \tau \leq t$, and $\xi_n(\tau) \uparrow \xi(\tau)$, as $n \to \infty$, for $0 \leq \tau \leq t$. Since $\sigma(\cdot,\tau)$ and $h(\cdot,\tau)$ are left-continuous, letting $n \to \infty$ in (3.22) we establish (3.7) when (3.21) holds. To verify (3.8) in the situation (3.21), it suffices to locate $\bar{\tau} \in (0,t)$ such that

$u(\xi(\bar{t})+,\bar{t}) = u(\xi(\bar{t}),\bar{t})$, to set $\bar{x} = \xi(\bar{t})$ and to note that the restriction of $\xi(\cdot)$ on $[0,\bar{t}]$ is the minimal characteristic emanating from a point $(\bar{x},\bar{t})$ at which $u(\bar{x}+,\bar{t}) = u(\bar{x},\bar{t})$. This completes the proof. ∎

4. Proof of Proposition 2.1

In the sequel, Λ will denote a generic positive constant that depends solely on T , i.e., it is independent of the initial data.

We define

(4.1) $$S(t) := \sup|\sigma(\cdot,t)| \;, \qquad 0 \leq t \leq T \,.$$

By virtue of (1.3), (1.4), and (2.4),

(4.2) $$|f(u(x,t))| \leq S(t) + \Lambda \int_0^t |f(u(x,\tau))|\,d\tau \,, \qquad -\infty < x < \infty \,, \quad 0 \leq t \leq T \;,$$

whence

(4.3) $$|f(u(x,t))| \leq S(t) + \Lambda \int_0^t S(\tau)\,d\tau \;, \qquad -\infty < x < \infty \,, \quad 0 \leq t \leq T \,.$$

Combining (3.6), (3.5), (2.4), and (4.3), we get

(4.4) $$|h(x,t)| \leq \Lambda S(t) + \Lambda \int_0^t S(\tau)\,d\tau \;, \qquad -\infty < x < \infty \,, \quad 0 \leq t \leq T \,.$$

We now fix a point (x,t) in $(-\infty,\infty) \times (0,T]$ and consider the minimal characteristic $\xi(\cdot)$ that emanates from it. By account of (3.7), (3.8),

(4.5) $$|\sigma(x,t)| \leq \sup|f(u_0(\cdot))| + \int_0^t |h(\xi(\tau),\tau)|\,d\tau \;, \qquad -\infty < x < \infty \,, \quad 0 \leq t \leq T \,,$$

and this, together with (4.1), (4.4), implies

(4.6) $$S(t) \leq \sup|f(u_0(\cdot))| + \Lambda \int_0^t S(\tau)\,d\tau \,, \qquad 0 \leq t \leq T \,,$$

whence

$$S(t) \leqslant \Lambda \sup|f(u_0(\cdot))| \quad , \qquad 0 \leqslant t \leqslant T . \tag{4.7}$$

Combining (4.7) with (4.3), (2.1), and (2.2), we deduce (2.8), for a constant M that depends solely on T and $\sup|u_0(\cdot)|$. This completes the proof. ∎

5. Proof of Proposition 2.2

Throughout this section, M will denote a generic positive constant that depends solely on $\sup|u_0(\cdot)|$ and T while N will stand for a generic positive constant that may depend on $\underset{[0,1]}{TV}\, u_0(\cdot)$ as well as on $\sup|u_0(\cdot)|$ and T .

Proof of Estimate (2.9). Let us set

$$F(t) := \underset{[0,1]}{TV}\, f(u(\cdot,t)) , \qquad 0 \leqslant t \leqslant T , \tag{5.1}$$

$$W(t) := \underset{[0,1]}{TV}\, \sigma(\cdot,t) , \qquad 0 \leqslant t \leqslant T . \tag{5.2}$$

Combining (1.3), (1.4), (2.1), and (2.8), we deduce

$$F(t) \leqslant W(t) + M\int_0^t F(\tau)\,d\tau, \qquad 0 \leqslant t \leqslant T , \tag{5.3}$$

whence

$$F(t) \leqslant W(t) + M\int_0^t W(\tau)\,d\tau, \qquad 0 \leqslant t \leqslant T . \tag{5.4}$$

By virtue of (3.6), (3.5) and upon using again (2.1), (2.8), it follows from (5.4)

$$\underset{[0,1]}{TV}\, h(\cdot,t) \leqslant M W(t) + M\int_0^t W(\tau)\,d\tau , \quad 0 \leqslant t \leqslant T . \tag{5.5}$$

We now fix $t \in (0,T]$ and take any mesh $0 = x_0 < x_1 < \cdots < x_n = 1$. Let $\xi_i(\cdot)$ denote the minimal characteristic emanating from (x_i,t) , $i = 0, \cdots , n$. From (3.7), (3.8) we get

$$(5.6)\qquad \sum_{i=1}^{n} |\sigma(x_i,t) - \sigma(x_{i-1},t)| \leq \underset{[0,1]}{TV} f(u_0(\cdot)) + \int_0^t \sum_{i=1}^{n} |h(\xi_i(\tau),\tau) - h(\xi_{i-1}(\tau),\tau)|\, d\tau$$

Combining (5.6), (5.2), and (5.5), we deduce

$$(5.7)\qquad W(t) \leq \underset{[0,1]}{TV} f(u_0(\cdot)) + M\int_0^t W(\tau)\,d\tau\ ,\quad 0 \leq t \leq T\ ,$$

whence it follows $W(t) \leq N$, $0 \leq t \leq T$, and, therefore, by (5.4), $F(t) \leq N$, $0 \leq t \leq T$. This, in turn, yields (2.9), by account of (5.1), (2.1), and (2.8). The proof is complete. ■

Proof of Estimate (2.10). We fix $x \in (-\infty,\infty)$, pick any mesh $0 = t_0 < t_1 < \cdots < t_n = T$ and let $\xi_i(\cdot)$ denote the minimal characteristic emanating from (x,t_i), $i = 1, \cdots, n$. Since $\xi_i(\tau)$ is contained in an *a priori* bounded interval $[x-M,x]$, $i = 1, \cdots, n$, it follows from (3.7), (3.8) that

$$(5.8)\qquad \sum_{i=1}^{n} |\sigma(x,t_i) - \sigma(x,t_{i-1})| \leq M \underset{[0,1]}{TV} f(u_0(\cdot)) + \sum_{i=1}^{n-1} \int_{t_{i-1}}^{t_i} \sum_{j=i+1}^{n} |h(\xi_j(\tau),\tau) - h(\xi_{j-1}(\tau),\tau)|\, d\tau + \sum_{i=1}^{n} \int_{t_{i-1}}^{t_i} |h(\xi_i(\tau),\tau)|\, d\tau\ .$$

By account of (3.6), (3.5), (2.8), and (2.9) we have $|h(x,t)| \leq M$, $-\infty < x < \infty$, $0 \leq t \leq T$, and $\underset{[x-M,x]}{TV} h(\cdot,t) \leq N$, $0 \leq t \leq T$, and so (5.8) yields

$$(5.9)\qquad \underset{[0,T]}{TV}\, \sigma(x,\cdot) \leq N\ ,\qquad -\infty < x < \infty\ .$$

Recalling (1.3), (1.4) and since $\underset{[0,T]}{TV} z(x,\cdot) \leq M$, (5.9) implies $\underset{[0,T]}{TV} f(u(x,\cdot)) \leq N$, $-\infty < x < \infty$, whence (2.10) follows, with the help of (2.1) and (2.8). This completes the proof. ■

Proof of Estimate (2.11). We fix $0 < s < t \leq T$. For each $x \in (-\infty,\infty)$ we consider the minimal characteristic $\xi(\cdot)$ emanating from (x,t) and set $y(x) := x - \xi(s)$. By (3.7),

$$\sigma(x,t) = \sigma(x-y(x),s) - \int_s^t h(\xi(\tau),\tau)\,d\tau \tag{5.10}$$

and so, recalling that $|h(x,\tau)| \leq M$, $-\infty < x < \infty$, $s \leq \tau \leq t$, $\underset{[0,1]}{TV}\sigma(\cdot,s) \leq N$, and noting that $0 < y(x) < M(t-s)$, $-\infty < x < \infty$, we deduce

$$\begin{aligned}\int_0^1 |\sigma(x,t) - \sigma(x,s)|\,dx &\leq \int_0^1 |\sigma(x-y(x),s) - \sigma(x,s)|\,dx + M(t-s) \\ &\leq \sup y(\cdot)\,\underset{[0,1]}{TV}\sigma(\cdot,s) + M(t-s) \leq N(t-s)\,.\end{aligned} \tag{5.11}$$

Combining (5.11) with (1.3) and recalling that $\partial_t z(x,t) = h(x,t)$ we get

$$\begin{aligned}\int_0^1 |f(u(x,t)) - f(u(x,s))|\,dx &\leq \int_0^1 |\sigma(x,t) - \sigma(x,s)|\,dx + \int_s^t\int_0^1 |h(x,\tau)|\,dx\,d\tau \\ &\leq N(t-s)\end{aligned} \tag{5.12}$$

whence (2.11) follows, with the help of (2.1), (2.8). The proof is complete. ■

Proof of Estimate (2.12). We fix $-\infty < x < y < \infty$. For each $t \in (0,T]$ we consider the minimal characteristic $\xi(\cdot)$ emanating from (y,t) and identify $r \in (0,T]$ with the property that when $t \in (0,r)$ then $\xi(0) \in (x,y)$ while if $t \in [r,T]$ then $\xi(0) \leq x$. It is clear that $r < M(y-x)$ and so

$$\int_0^r |\sigma(y,t) - \sigma(x,t)|\,dt \leq M(y-x)\,. \tag{5.13}$$

For each fixed $t \in [r,T]$, we introduce the inverse function $\phi(\cdot)$ of $\xi(\cdot)$ and set $s(t) := t - \phi(x)$. Thus $s(t)$ measures the time it takes $\xi(\cdot)$ to traverse the interval $[x,y]$. In particular,

$$0 < s(t) < M(y-x)\,, \qquad r \leq t \leq T\,. \tag{5.14}$$

By virtue of (3.7) and (3.2),

$$\sigma(y,t) = \sigma(x,t-s(t)) - \int_x^y \frac{h(\xi,\phi(\xi))}{f'(u(\xi,\phi(\xi)))}\,d\xi\,, \qquad r \leqslant t \leqslant T\,, \tag{5.15}$$

and so, by account of (2.1), (2.8),

$$\begin{aligned}\int_r^T |\sigma(y,t) - \sigma(x,t)|\,dt &\leqslant \int_r^T |\sigma(x,t-s(t)) - \sigma(x,t)|\,dt + M(y-x) \\ &\leqslant \sup s(\cdot)\, \underset{[0,T]}{TV}\, \sigma(x,\cdot) + M(y-x)\,.\end{aligned} \tag{5.16}$$

Therefore, combining (5.13), (5.16), (5.14), and (5.9) we conclude

$$\int_0^T |\sigma(y,t) - \sigma(x,t)|\,dt \leqslant N(y-x)\,. \tag{5.17}$$

By virtue of (2.1), we get from (1.3), (1.4)

$$\begin{aligned}f(u(y,t)) - f(u(x,t)) &= \sigma(y,t) - \sigma(x,t) \\ &+ \int_0^t K(t-\tau)\left\{g\circ f^{-1}\circ f(u(y,\tau)) - g\circ f^{-1}\circ f(u(x,\tau))\right\}d\tau\end{aligned} \tag{5.18}$$

and so

$$|f(u(y,t)) - f(u(x,t))| \leqslant |\sigma(y,t) - \sigma(x,t)| + M\int_0^t |f(u(y,\tau)) - f(u(x,\tau))|\,d\tau \tag{5.19}$$

whence

$$|f(u(y,t)) - f(u(x,t))| \leqslant |\sigma(y,t) - \sigma(x,t)| + M\int_0^t |\sigma(y,\tau) - \sigma(x,\tau)|\,d\tau\,. \tag{5.20}$$

From (5.20) and (5.17) we deduce

$$\int_0^T |f(u(y,t)) - f(u(x,t))|\,dt \leqslant N(y-x) \tag{5.21}$$

and (2.12) follows, with the help of (2.1) and (2.8). This completes the proof. ∎

6. Proof of Proposition 2.3

Throughout this section we assume that (2.13), (2.14), and (2.15) hold, in addition to (2.1), (2.2), (2.3), and (2.4). As in earlier sections, M will denote a generic positive constant that may depend at most on T and $\sup|u_0(\cdot)|$ and so, in particular, it is independent of $TVu_0(\cdot)$.

Lemma 6.1. *It is*

(6.1) $$\underset{[0,1]}{TV}\, z(\cdot,t) \leqslant M\ , \qquad 0 \leqslant t \leqslant T\ ,$$

(6.2) $$\underset{[0,1]}{TV}\, p(\cdot,t) \leqslant M\ , \qquad 0 \leqslant t \leqslant T\ .$$

Proof. Assuming, for definiteness, $g'(u)/f'(u)$ is nonincreasing, we define the convex function

(6.3) $$\eta(u) := -\int_0^u \frac{g'(v)}{f'(v)}\, dv\ , \qquad -\infty < u < \infty\ .$$

We claim that the "entropy" inequality

(6.4) $$\partial_t\eta(u(x,t)) - \partial_x g(u(x,t)) - \eta'(u(x,t))\partial_x z(x,t) \leqslant 0$$

is satisfied by the solution $u(x,t)$, in the sense of measures. Indeed, it is clear that (6.4) holds, as an equality, whenever $u(x,t)$ is a Lipschitz solution of (1.1), (1.3). Therefore, for BV solutions considered here, the left-hand side of (6.4) is a measure supported on the set of points of approximate jump discontinuity of $u(x,t)$. Hence, to get (6.4) it suffices to verify

(6.5) $$s\{\eta(u(x+,t)) - \eta(u(x,t))\} + g(u(x+,t)) - g(u(x,t)) \geqslant 0$$

at points of discontinuity, where s denotes the speed of propagation of the discontinuity as determined via the Rankine-Hugoniot jump condition

$$(6.6)\qquad -s\{u(x+,t) - u(x,t)\} + f(u(x+,t)) - f(u(x,t)) = 0 .$$

A simple and familiar calculation shows that (6.6) implies (6.5), by virtue of the convexity of $\eta(u)$, (2.3) and (2.7).

We form the convolution of (6.4) with $K(\cdot)$, recalling (2.14), and use (1.4) to get

$$(6.7)\qquad \begin{aligned} \partial_x z(x,t) &\geq -\int_0^t K(t-\tau)\eta'(u(x,\tau))\,\partial_x z(x,\tau)\,d\tau \\ &\quad + K(0)\eta(u(x,t)) - K(t)\eta(u_0(x)) + \int_0^t K'(t-\tau)\eta(u(x,\tau))\,d\tau \\ &\geq -M\int_0^t |\partial_x z(x,\tau)|\,d\tau - M . \end{aligned}$$

Similarly, taking the convolution of (6.4) with $K'(\cdot)$ and using (2.15) and (3.5) we deduce

$$(6.8)\qquad \begin{aligned} \partial_x p(x,t) &\leq -\int_0^t K'(t-\tau)\eta'(u(x,\tau))\,\partial_x z(x,\tau)\,d\tau \\ &\quad + K'(0)\eta(u(x,t)) - K'(t)\eta(u_0(x)) + \int_0^t K''(t-\tau)\eta(u(x,\tau))\,d\tau \\ &\leq M\int_0^t |\partial_x z(x,\tau)|\,d\tau + M . \end{aligned}$$

Letting IV denote increasing variation and DV stand for decreasing variation, we get from (6.7) and (6.8)

$$(6.9)\qquad \underset{[0,1]}{TV}\, z(\cdot,t) = 2\,\underset{[0,1]}{DV}\, z(\cdot,t) \leq M\int_0^t \underset{[0,1]}{TV}\, z(\cdot,\tau)\,d\tau + M ,$$

$$(6.10)\qquad \underset{[0,1]}{TV}\, p(\cdot,t) = 2\,\underset{[0,1]}{IV}\, p(\cdot,t) \leq M\int_0^t \underset{[0,1]}{TV}\, z(\cdot,\tau)\,d\tau + M .$$

From (6.9) we deduce (6.1) and then, using (6.10), we obtain (6.2). This completes the proof. ∎

Proof of Estimate (2.16). We fix $t \in (0,T]$ and take any mesh $0 \leq x_1 < x_2 \leq \cdots \leq x_{2i-1} < x_{2i} \leq \cdots \leq x_{2n-1} < x_{2n} \leq 1$ such that

(6.11) $$\sigma(x_{2i-1},t) \leq \sigma(x_{2i},t) , \qquad i = 1, \cdots, n .$$

Let $\xi_i(\cdot)$ denote the minimal characteristic emanating from (x_i,t). We use (3.2), (2.1), the mean value theorem and then (1.3) to get

(6.12)
$$\begin{aligned} x_{2i} - x_{2i-1} &= \xi_{2i}(0) - \xi_{2i-1}(0) + \int_0^t \{f'(u(\xi_{2i}(\tau),\tau)) - f'(u(\xi_{2i-1}(\tau),\tau))\} d\tau \\ &\geq \int_0^t \{f' \circ f^{-1} \circ f(u(\xi_{2i}(\tau),\tau)) - f' \circ f^{-1} \circ f(u(\xi_{2i-1}(\tau),\tau))\} d\tau \\ &= \int_0^t P_i(\tau)\{f(u(\xi_{2i}(\tau),\tau)) - f(u(\xi_{2i-1}(\tau),\tau))\} d\tau \\ &= \int_0^t P_i(\tau)\{\sigma(\xi_{2i}(\tau),\tau) - \sigma(\xi_{2i-1}(\tau),\tau)\} d\tau \\ &\quad + \int_0^t P_i(\tau)\{z(\xi_{2i}(\tau),\tau) - z(\xi_{2i-1}(\tau),\tau)\} d\tau \end{aligned}$$

with

(6.13) $$P_i(\tau) = \frac{f''(u_i(\tau))}{f'(u_i(\tau))}$$

where $u_i(\tau)$ is a number between $u(\xi_{2i}(\tau),\tau)$ and $u(\xi_{2i-1}(\tau),\tau)$. In particular, on account of (2.3), (2.1), and (2.8),

(6.14) $$0 < \mu \leq P_i(\tau) \leq M , \qquad 0 \leq \tau \leq t , \qquad i = 1, \cdots, n .$$

Furthermore, using (3.7) and (3.6),

(6.15)
$$\begin{aligned} \sigma(\xi_{2i}(\tau),\tau) - \sigma(\xi_{2i-1}(\tau),\tau) &= \sigma(x_{2i},t) - \sigma(x_{2i-1},t) \\ &\quad + K(0) \int_\tau^t \{g(u(\xi_{2i}(s),s)) - g(u(\xi_{2i-1}(s),s))\} ds \\ &\quad + \int_\tau^t \{p(\xi_{2i}(s),s) - p(\xi_{2i-1}(s),s)\} ds . \end{aligned}$$

Combining (6.12), (6.15), (6.14), and (6.11) we conclude

$$x_{2i} - x_{2i-1} \geq \mu t\left\{\sigma(x_{2i},t) - \sigma(x_{2i-1},t)\right\} \tag{6.16}$$

$$+ K(0) \int_0^t P_i(\tau) \int_\tau^t \left\{g(u(\xi_{2i}(s),s)) - g(u(\xi_{2i-1}(s),s))\right\} ds\, d\tau$$

$$+ \int_0^t P_i(\tau) \int_\tau^t \left\{p(\xi_{2i}(s),s) - p(\xi_{2i-1}(s),s)\right\} ds\, d\tau .$$

We note that

$$\underset{[0,1]}{TV}\, \sigma(\cdot,t) = 2 \underset{[0,1]}{IV}\, \sigma(\cdot,t) = 2 \sup \sum_i \left\{\sigma(x_{2i},t) - \sigma(x_{2i-1},t)\right\} \tag{6.17}$$

where the supremum on the right-hand side of (6.17) is taken over all meshes $0 \leq x_1 < \cdots \leq x_{2i-1} < x_{2i} \leq \cdots < x_{2n} \leq 1$ which satisfy (6.11). Therefore, (6.16) together with (6.2) induce the estimate

$$t \underset{[0,1]}{TV}\, \sigma(\cdot,t) \leq M \int_0^t \int_\tau^t \underset{[0,1]}{TV}\, g(u(\cdot,s))\, ds\, d\tau + M . \tag{6.18}$$

On the other hand, using (2.1), (2.8), and then (1.3), (6.1), we get

$$\underset{[0,1]}{TV}\, g(u(\cdot,s)) \leq M \underset{[0,1]}{TV}\, f(u(\cdot,s)) \leq M \underset{[0,1]}{TV}\, \sigma(\cdot,s) + M \tag{6.19}$$

and so (6.18) yields

$$t \underset{[0,1]}{TV}\, \sigma(\cdot,t) \leq M \int_0^t \int_\tau^t \underset{[0,1]}{TV}\, \sigma(\cdot,s)\, ds\, d\tau + M \tag{6.20}$$

$$= M \int_0^t \tau \underset{[0,1]}{TV}\, \sigma(\cdot,\tau)\, d\tau + M$$

whence

$$\underset{[0,1]}{TV}\, \sigma(\cdot,t) \leq \frac{M}{t} , \qquad 0 < t \leq T . \tag{6.21}$$

It is now clear that (2.16) follows from (6.21), (1.3), (6.1), (2.1), and (2.8). This completes the proof. ∎

References

[1] Coleman, B. D., and Gurtin, M. E., Waves in materials with memory, II. On the growth and decay of one-dimensional acceleration waves. Arch. Rational Mech. Anal. 19 (1965), 266-298.

[2] Dafermos, C. M., Generalized characteristics and the structure of solutions of hyperbolic conservation laws. Indiana U. Math. J. 26 (1977), 1097-1119.

[3] Dafermos, C. M., Dissipation in materials with memory. *Viscoelasticity and Rheology*, pp. 221-234 (A. Lodge, J. A. Nohel, and M. Renardy, eds.) Academic Press 1985.

[4] Dafermos, C. M., Development of singularities in the motion of materials with fading memory. Arch. Rational Mech. Anal. 91 (1986), 193-205.

[5] Dafermos, C. M., Solutions in L^∞ for a conservation law with memory. (To appear.)

[6] Dafermos, C. M., and Nohel, J. A., A nonlinear hyperbolic Volterra equation in viscoelasticity. Am. J. Math. Suppl. dedicated to P. Hartman (1981), 87-116.

[7] DiPerna, R. J., Convergence of approximate solutions to conservation laws. Arch. Rational Mech. Anal. 82 (1983), 27-70.

[8] Filippov, A. F., Differential equations with discontinuous right-hand side. Mat. Sbornik (N. S.) 51 (93) (1960), 99-128.

[9] Glimm, J., Solutions in the large for nonlinear hyperbolic systems of equations. Comm. Pure Appl. Math. 18 (1965), 697-715.

[10] Glimm, J., and Lax, P. D., Decay of solutions of systems of nonlinear hyperbolic conservation laws. Memoirs A. M. S. 101 (1970).

[11] Greenberg, J. M., The existence and qualitative properties of solutions of $\partial_t u + \frac{1}{2}\partial_x(u^2 + \int_0^t c(s)\, u^2(x,t-s)ds) = 0$. J. Math. Anal. Appl. 42 (1973), 205-220.

[12] Hrusa, J. W., and Nohel, J. A., The Cauchy problem in one-dimensional nonlinear viscoelasticity. J. Diff. Eqs. 59 (1985), 388-412.

[13] Klainerman, S., and Majda, A., Formation of singularities for wave equations including the nonlinear vibrating string. Comm. Pure Appl. Math. 33 (1980), 241-263.

[14] Lax, P. D., Development of singularities of solutions of nonlinear hyperbolic partial differential equations. J. Math. Physics 5 (1964), 611-613.

[15] MacCamy, R. C., A model for one-dimensional, nonlinear viscoelasticity. Quart. Appl. Math. 35 (1977), 21-33.

[16] Malek-Madani, R., and Nohel, J. A., Formation of singularities for a conservation law with memory. SIAM J. Math. Anal. 16 (1985), 530-540.

[17] Nohel, J. A., A nonlinear conservation law with memory. *Volterra and Functional Differential Equations*, pp. 91-123 (K. B. Hannsgen, T. L. Herdman and R. L. Wheeler, eds.) Marcel Dekker 1982.

[18] Nohel, J. A., and Renardy, M., Development of singularities in nonlinear viscoelasticity. This volume.

[19] Oleinik, O. A., Discontinuous solutions of non-linear differential equations. Usp. Mat. Nauk (N. S.) 12 (3) (1957), 3-73.

[20] Rascle, M., Un resultat de "compacite par compensation a coefficients variables." Application a l'elasticite nonlineaire. Compt. Rend. Acad. Sci. Paris, Serie I, 302 (1986), 311-314.

[21] Renardy, M., Recent developments and open problems in the mathematical theory of viscoelasticity. *Viscoelasticity and Rheology*, pp. 345-360 (A. Lodge, J. A. Nohel, and M. Renardy, eds.) Academic Press 1985.

[22] Truesdell, C. A., and Noll, W., *The Nonlinear Field Theories of Mechanics*. Handbuch der Physik III/3. Springer-Verlag, Berlin 1965.

HYPERBOLIC DYNAMICS IN THE FLOW OF ELASTIC LIQUIDS

D. D. Joseph
Department of Aerospace Engineering and Mechanics
The University of Minnesota
Minneapolis, Minnesota 55455

Summary

In this paper I discuss concepts of viscosity, elasticity, hyperbolicity, Hadamard instability and change of type in the flow of viscoelastic fluids

1. Constitutive Equations

Constitutive equations relate stress and deformation. Too many constitutive equations have been proposed by people to get one model which will describe all the possible motions of a fluid. Since the variety of responses which are available to viscoelastic fluids is very great, a single equation which accounts for everything may be too abstract to be of much practical use. Eqs. (1.10) and (1.12) are examples of too abstract equations. More specific models are useful only when the domain of deformations in which they live is specified. Therefore, in an ideal world we could have a model valid within a prescribed class of deformations.

There is a great simplification in the problem of constitutive modeling when the deformations are a small perturbation of states of rest. These deformations depend on a Newtonian viscosity μ and a smooth relaxation function $G(s)$, where $G(s) > 0$, $G'(s) < 0$ for $0 \leq s=t-\tau \leq \infty$, and τ is the past time. The stress $\boldsymbol{\tau}$ is given by

$$\boldsymbol{\tau} = 2\mu\mathbf{D}[\mathbf{u}] + 2\int_0^\infty G(s)\mathbf{D}[\mathbf{u}(\mathbf{x},t-s)]\, ds \tag{1.1}$$

where μ is the Newtonian viscosity, $\mathbf{u}(\mathbf{x},t)$ is the velocity and $\mathbf{D}[\mathbf{u}]$ the symmetric part of the velocity gradient. Equation (1.1) is a Jeffreys' type of generalization of Boltzmann's equation of linear viscoelasticity in which the presence of a Newtonian contribution is acknowledged. Eq. (1.1) also holds in the class of small perturbations of rigid motions. We might think that (1.1) models a polymeric solution in which the solvent is Newtonian and the polymers add elasticity.

A constitutive equation of the rate type may be obtained as the time derivative of (1.1)

$$\frac{\partial \boldsymbol{\tau}}{\partial t} = 2\mu \frac{\partial \mathbf{D}}{\partial t} + 2G(0)\mathbf{D} + 2\int_0^\infty G'(s)\mathbf{D}[\mathbf{u}(\mathbf{x},t-s)]ds. \qquad (1.2)$$

Jeffreys' model is a special case of (1.2) in which

$$G(s) = \frac{\eta}{\lambda} e^{-s/\lambda} \qquad (1.3)$$

where λ is the relaxation time and η is the elastic viscosity. Combining (1.2) and (1.3) we get

$$\lambda \frac{\partial \boldsymbol{\tau}}{\partial t} = 2\mu\lambda \frac{\partial \mathbf{D}}{\partial t} + 2\tilde{\mu}\mathbf{D} - \boldsymbol{\tau}. \qquad (1.4)$$

A retardation time

$$\Lambda = \mu\lambda/\eta \qquad (1.5)$$

is usually defined for (1.4). When $\mu = 0$, (1.4) gives rise to a Maxwell model

$$\lambda \frac{\partial \boldsymbol{\tau}}{\partial t} = 2\eta\mathbf{D} - \boldsymbol{\tau}. \qquad (1.6)$$

Fluids with $\mu = 0$ are like relaxing elastic solids. They propagate shock waves. Fluids with $\mu \neq 0$ are diffusive; they smooth shocks.

Equations (1.1) and (1.2) are perturbation equations and are naturally not invariant under changes of frame which do not satisfy the same conditions of linearization. Various invariant theories which are said to be linear have been proposed. For example Coleman and Noll [1961] linearized a functional depending on the history of the right relative Cauchy-Green strain tensor. Naturally they arrive at a linear expansion, linear in this non-linear tensor. They call this "the finite linear theory of viscoelasticity." When applied to incompressible fluids they get (1.1) with $\mu = 0$ and $\mathbf{D}$ replaced with the s derivative of $C_t(\mathbf{x},t-s) - \mathbf{1} = \mathbf{G}(s)$, $\mathbf{G}(0) = 0$. The linearization of $G(s)$ around $\mathbf{0}$ is $\mathbf{D}$. If the kernel $G(s) = \frac{\eta}{\lambda} e^{-s/\lambda}$ is of Maxwell's type, then Coleman and Noll's equation is a lower convected Maxwell model. If we suppose that the stress functional depends on the Finger tensor, rather than the Cauchy tensor, we arrive at Lodge's theory, which is the same as an upper convected Maxwell model when the kernel is of Maxwell's type. Saut and Joseph [1985] under different hypotheses than Coleman and Noll arrived at (1.1) with G(s) in the place of **D** under the integral and $\mu \neq 0$. If Saut and Joseph had used $H(s) = C_t^{-1}(x,t-s) - 1$ instead of G(s) they would have H(s) replacing **D** under the integral. The rate equations for an equation of the Saut-Joseph type, with a kernel of Maxwell type, is an Oldroyd B. None of these so called linear equations are completely linearized. When they are completely linearized they reduce universally to (1.1) and (1.2). These two equations are model independent. They apply to all viscoelastic fluids in motions which perturb rest. This shows that the Newtonian viscosity μ and the relaxation function G(s) are genuine material parameters which are also model independent.

To our knowledge, the first person to introduce a rate equation with a Newtonian viscosity and relaxing elasticity was H. Jeffreys (1929, p. 265). Most of the models arising from molecular modeling of polymeric solutions, like those of Rouse and Zimm, have a Newtonian contribution from the solvent and are of the Jeffreys' type. An invariant formulation of rate equations containing relaxation and retardation (Newtonian viscosity) effects evidently first appears in the celebrated 1950 paper of Oldroyd. Green and Rivlin (1960) appear to have been the first to introduce Newtonian viscosity to integral models. They get rate terms from integrals by allowing delta functions and their derivatives in the kernels. Saut and Joseph (1983) derived integral expressions of the type introduced by Green and Rivlin from a theory of fading memory in which the ensemble of all possible linearized stresses coincides with certain topological dual of a domain space (say, a Sobolev space) for allowed deformations. Maxwell models and the generalization of these embodied in the theory of fading memory of Coleman and Noll (see SJ for references) cannot contain a Newtonian viscosity. These models are all instantaneously elastic. Various kinds of hyperbolic phenomena, waves, shock waves, loss of evolution, Hadamard instabilities, change of type arise in fluids with instantaneous elasticity (see Joseph, Renardy and Saut (1985), Joseph (1985A), Joseph and Saut (1985) and this review). Many distinguished scientists of the 19th and early 20th century, Poisson, Maxwell, Poynting, Boltzmann believed that liquids were closer to solids than to gases, with instantaneous and relaxing elasticity, and there is also a line of interesting experiments of this same period which explore this idea (see Joseph, 1985B for a recent historical perspective). The results given in this paper are an entry in this history. The notion of instantaneous and relaxing elasticity can be reconciled with polymers in Newtonian solvents by supposing that the solvents are elastic and not Newtonian.

We have argued that the response to motions perturbing rest (or rigid motions) is completely determined when the Newtonian viscosity μ and the relaxation function G(s) are known. G(s) gives the relaxation for $s \geq 0$ of stresses after a sudden step in displacement. The name "shear stress modulus" or "shear modulus" or "elastic modulus" will be reserved for the largest value G(0) of G(s).

It is helpful that we understand viscosity in the following way. Suppose that we are in the case of steady shearing with one component of velocity u(x) depending on one variable x. The shear stress $\tau(\kappa) = \tau_{12}$ of $\boldsymbol{\tau}$ depend then on the rate of shear $\kappa(x) = D_{12}$ of **D** and (1.1) reduces to $\tau = (\mu + \eta)\kappa$ where

$$\tilde{\mu} = \mu + \eta \tag{1.7}$$

is the zero shear viscosity and

$$\eta = \int_0^\infty G(s)\, ds \tag{1.8}$$

is the elastic viscosity. Newtonian fluids have $\eta = 0$, $\tilde{\mu} = \mu$. Elastic fluids have $\mu = 0$, $\tilde{\mu} = \eta$. In general

$$\tilde{\mu} \geq \eta \tag{1.9}$$

with equality for elastic fluids. It is easy to measure the zero shear viscosity $\tilde{\mu}$, but the measurement leaves μ and η undetermined. Elastic fluids ($\mu = 0$) with short memories can appear to be Newtonian in standard rheometrical tests.

We turn now to the problem of constitutive modeling in the general case. A framework for such modeling could start from Noll's theory of a simple incompressible fluid. In this, the stress at a particle **x** is given by a functional on the history of right relative Cauchy tensor $C_t(\tau)$ or the history of $\mathbf{G}(s) = C_t(\tau) - 1$; $\boldsymbol{\tau} = \mathbf{F}[\mathbf{G}]$. To assign meaning to **F**[ ] it is necessary to specify its domain. The Coleman-Noll theory of fading memory is a consequence of the assumption **G**(s) lies in a weighted $L^2_h(0,\infty)$ Hilbert space with weight h which makes the large s values of **G**(s) irrelevant. However, this large domain excludes some well established material models and phenomena. Saut and Joseph (1983) showed that by restricting the allowed domain of **F**, the topological dual may be enlarged leading to distributions in the dual and rate terms in the constitutive model. Another method, presented here, is to keep the large domain with **G**(s) $L^2_h(0,\infty)$ but to add **f**(**D**) where **f** is an ordinary isotropic symmetric tensor valued function of **D**(**x**,t). Thus

$$\boldsymbol{\tau} = \mathbf{f}(\mathbf{D}) + \mathop{\mathbf{F}}_{s=0}^{\infty}[\mathbf{G}(s)] \; . \tag{1.10}$$

In this decomposition **f**(**D**) is the viscous part, and **F** the elastic. It may be assumed without losing much generality that **f**(**D**) is a quadratic polynomial in D with coefficients which depend on the invariant scalars of **D**. Equation (1.1) follows easily form linearizing (1.10) on states of rest, by representing linear functionals with scalar products (integrals) in the weighted $L^2(0,\infty)$ the domain space of G(s).

Joseph, Renardy and Saut (1985), hereafter called JRS, derived the general form of the constitutive equation of rate type for any elastic liquid in any motion. This equation arises from calculations following the time differentiation of the stress functional and it may be written as

$$\frac{d\tau_{ij}}{dt} = S_{ijkp}D_{kp}[u] + A_{ijkp}\Omega_{kp}[u] + N_{ij} \tag{1.11}$$

where all tensors are symmetric in (i,j), D_{kp} is the symmetric part of the velocity gradient $\partial u_k/\partial x_p$ and Ω_{kp} is the skew symmetric part. The fourth order tensors **S** and **A** are expressible by integrals and N_{ij} is of "lower order" in the sense of hyperbolic analysis. More discussion of this equation can be found in the paper of Joseph (1985).

When the liquid also has a viscous response we may replace (1.11) by

$$\frac{d\tau_{ij}}{dt} = R_{ijkp}\frac{dD_{ij}}{dt} + S_{ijkp}D_{kp} + A_{ijkp}\Omega_{kp} + N_{ij} \tag{1.12}$$

where **R** is a fourth order tensor valued function $R_{ijk\ell} = \partial f_{ij}/\partial D_{k\ell}$. In the fully linearized case, $R_{ijk\ell} = 2\mu\delta_{ik}\delta_{j\ell}$. JRS identified a class of models which are more general than (1.11) but contain all the special models of rate equations which appear in the rheological literature. In their model, the symmetric fourth order tensor **S** is expressed as the most general form involving any second order tensor **P**. When this is applied to (1.12) we find that

$$\frac{d\tau}{dt} = R\frac{dD}{dt} + \mathbf{P}\cdot\mathbf{D} + (\mathbf{P}\cdot\mathbf{D})^T + \frac{1}{2}[\mathbf{A}\cdot\mathbf{\Omega} + (\mathbf{A}\cdot\mathbf{\Omega})^T] + \mathbf{N} \tag{1.13}$$

No assumption is made about the fourth order tensor **A**. The special models which are studied in the rheological literature are such that the tensors **P**, **A** and **N** are expressible in terms of the extra stress $\boldsymbol{\tau}$. These models include those of Oldroyd, Maxwell, Giesekus, Leonev, Phan Thien and Tanner and many others (see Joseph, (1985)).

When R = 0 the dynamical system associated with (1.13) supports hyperbolic waves of vorticity provided that the stresses do not enter a forbidden region in which the Cauchy problem is no longer well posed. These and some other phenomena are related to the type of a partial differential equation.

(1) The unsteady quasilinear problem is called evolutionary if, roughly speaking, the Cauchy problem for it is well posed (this is a notion strictly weaker than

hyperbolicity). The loss of evolution is an instability of the Hadamard type in which short waves will sharply increase in amplitude. For many models, those treated here, the problem of evolution may be conveniently framed in terms of vorticity.

(2) The steady quasilinear system may be analyzed for type. It is neither elliptic nor hyperbolic. On the other hand, the vorticity is either hyperbolic or elliptic, and it may change type, hyperbolic in some regions of flow and elliptic in others, as in transonic flow. We shall show that the full unsteady quasilinear system will undergo a loss of stability in the sense of Hadamard when the steady vorticity equation for inertialess flow is hyperbolic.

We consider a number of examples. Some models are always evolutionary and do not change type in unsteady flow. The vorticity equation for steady flow of such models can and does change type. Other models can become non-evolutionary and therefore undergo Hadamard instability. Some flows of these models are evolutionary, e.g.; shearing flows, while others are not evolutionary. In either case, the steady problem can undergo a change of type. Loss of evolution is impossible in flow perturbing uniform motion. To lose evolution it is necessary that certain stresses should exceed critical values. In this sense the loss of evolution can be identified with the problem of failure of numerical simulations at high Weissenberg numbers.

2. Loss of Evolution

The loss of evolution is a concept associated with the well posedness of the Cauchy problem. Let us consider a quasilinear system of the form

$$\mathbf{A}\,\frac{\partial \mathbf{u}}{\partial t} + \sum_{j=1}^{n} \mathbf{B}_j\,\frac{\partial \mathbf{u}}{\partial x_j} + \mathbf{b} = 0\,, \tag{2.1}$$

where $\mathbf{A}$, $\mathbf{B}_1$, ..., $\mathbf{B}_n$ are m x m matrix valued functions and $\mathbf{b}$ is an m vector depending on $\mathbf{u},\mathbf{x},t$. The system (2.1) is evolutionary in some domain D of $\mathbb{R}^m \times \mathbb{R}^n \times \mathbb{R}$ in the t direction if for every fixed $\mathbf{u},\mathbf{x},t$ in D and any unit n-vector $\boldsymbol{\nu}$, the eigenvalue problem

$$\left[-\lambda\mathbf{A} + \sum_{j=1}^{n} \nu_j\,\mathbf{B}_j\right]\mathbf{v} = 0 \tag{2.2}$$

has only real eigenvalues. The system (2.1) is hyperbolic in the t direction if (2.2) has m real eigenvalues, not necessarily distinct, and a set of m linearly independent eigenvectors.

To justify these definitions, let us consider the simple case where $\mathbf{b} = 0$ and the matrices $\mathbf{A}$, $\mathbf{B}_1$, ..., $\mathbf{B}_m$ are independent of $\mathbf{u},\mathbf{x},t$. Let $\boldsymbol{\nu}$ be a unit vector in $\mathbb{R}^n$. If (2.1) is evolutionary, then any plane wave solution of (2.1) propagating in the $\boldsymbol{\nu}$-direction $\mathbf{v}(\mathbf{x},t) = \mathbf{v}\exp(ik(\boldsymbol{\nu}\cdot\mathbf{x} - \lambda t))$, k real, has necessarily λ real. This prevents the so called Hadamard instability, i.e., the fact that at any time t the amplitude of $\mathbf{u}$ could become arbitrarily large, even if $\mathbf{u}$ is bounded (but highly oscillatory) at time $t = 0$. In this context hyperbolicity means that in every direction in space, m independent plane waves can propagate.

The quasilinear systems for the velocity, stresses and pressure of fluids with instantaneous elasticity are not hyperbolic in the usual sense. For these systems it is proper to think of the loss of evolution. On the other hand, the unsteady equation for the vorticity is hyperbolic when it is evolutionary and is evolutionary when hyperbolic.

The use of plane waves to study the well posedness of the Cauchy problem is justified, in general, for waves so short that $\mathbf{A}$, $\mathbf{B}_j$ and $\mathbf{b}$ have constant components in a small period cell defined by the wave.

Hadamard instabilities are much stronger than those studied in bifurcation theory. One cannot expect a secondary flow whose dynamics would be governed by the evolution of one or several modes. Rather, if the initial value problem becomes ill-posed, there are flow fields which would not occur even as transient states, since "random" disturbances containing all modes would blow up instantly. The importance of loss of evolution in the flow of viscoelastic fluids was first recognized by Rutkevitch [1969, 1970, 1972]. Many other results were given by Joseph, Renardy and Saut [1985], called JRS. Independent results, following in part out of discussions leading to the paper by JRS, were given by Ahrens, Joseph, Renardy and Renardy [1984], and Renardy [1984, 1985]. Dupret and Marchal [1984, 1985] have also given some new results on the problem of loss of evolution in three dimensional problems. It has been suggested that models of viscoelastic fluids which lose evolution are no good and should be discarded; this idea is wrong, the loss of evolution can actually warn us about certain physical instabilities like melt fracture. On the other hand, it is certain that the loss of evolution will produce a disaster in numerical simulations.

Some systems of equations never lose evolution. This is true of the dynamical system generated by Newtonian fluids and by upper and lower convected Maxwell models. These systems are not always closer to physics than systems which can lose evolution.

The loss of evolution is associated with the initial value problem. It can occur for some problems and not for others. This is the case for incompressible inviscid fluids governed by Euler's equations in two space dimensions. It is known that when the initial data are smooth this equation has a unique smooth solution defined for all times. However, certain well known problems associated with Euler's equations can become non-evolutionary. H. Aref [1985] has commented on this:

> "Historically, this issue seems first to have arisen in the context of the Kelvin-Helmholtz problem. When the "common" vortex sheet rolls up, it apparently becomes singular after a finite time. One can gauge this feature already from the vantage point of linearized stability analysis: In the Kelvin-Helmholtz problem the amplification rate of a wave of wavelength λ is inversely proportional to λ. Thus, short waves amplify faster than longer waves at all wavelengths, and so, except for some delay in actually exciting the short waves, the outcome is almost inevitably headed for a singularity of some kind. The ill-posedness of the problem consists in the loss of a certain degree of analyticity after a finite time. (This is sometimes called ill-posedness in the sense of Hadamard.)
>
> This feature in itself is of interest, at least mathematically. One may ask, for example, what kind of "weak solution(s)" to the problem are available after the singularity time? The physical significance of such solutions (should they arise in some reasonable systematic way) is not at all clear at present ... .
>
> From a more pragmatic point of view one must regard the emergence of a singularity as physically unacceptable, a feature that shows an inadequacy in the description of the problem. Clearly, the basic equations must be augmented in some way that will remove the singularity. The notion that perturbations of arbitrarily short wavelength grow arbitrarily fast cannot be a physically meaningful statement within the framework of a hydrodynamic theory. And, sure enough, any physically motivated mechanism that provides a cut-off at small length scales will also lend a regularizing aspect to the solution. Sometimes, indeed, the problem can (apparently) be completely regularized so that smooth solutions exist for all time given smooth initial data. Regularizing mechanisms of this kind include diffusion, interfacial tension and the introduction of a small but finite thickness of the interfacial transition region itself."

The famous fingering instability of Saffman and Taylor [1958] is an instability to short waves of the Kelvin-Helmholtz type. When the less viscous fluid in the saturated porous materials fingers into the more viscous liquid, the short waves are the most unstable; the system based on Darcy's law is not evolutionary. We do not advocate throwing out Darcy's law because of this.

It is known that some popular hydrodynamic models of flowing composites, used in the study of fluidized beds and for other applications arising in mixture theories, lead to ill posed initial value problems (Gidaspow, 1974; Lyczkowski, et al. 1978).

An interesting discussion of the loss of evolution in the equation of magnetohydrodynamics with the Hall effect taken into account is given by Brushlinskii and Morozov [1968]. The equations of magnetohydrodynamics for a nondissipative plasma plane flow will not be evolutionary if the Hall effect is taken into account. The loss of evolution seems to be associated with certain physical instabilities.

On the other hand, Kulikovskii and Regirer [1968] have shown that electrodynamic equations which change type in steady flow can lose evolution and go unstable. Such solutions therefore cannot be realized. In the words of Kulikovskii and Regirer

"Owing to the rapid increase of perturbations, nonevolutionary equations cannot describe correctly changes of any physical quantity in time. Nonevolutionary solutions of the nonlinear equations in many cases can be regarded as an oversimplification in the derivation of these equations by discarding terms which are small for evolutionary solutions, but they can be essential for the perturbations which display a rapid increase. As the short wave disturbances increase most rapidly, then these could be the terms containing space derivatives of higher order or mixed derivatives with respect to space or time."

A similar point of view was adopted by Rutkevitch [1968] in his discussion of loss of evolution in viscoelastic fluids. He says

"In order to describe the development of small perturbations in the region where evolutionarity of the initial conditions is not possible, the effect of supplementary physical parameters should be taken into account. In a real system, these parameters can be extremely small, but they play a definite role in establishing a finite upper limit for the rate of buildup of perturbations."

One such parameter, used now extensively in numerical simulations of non-Newtonian flows, is to add a Newtonian viscosity. In Section 8 we shall address this question from the point of view of rheometrical science.

The paper by Ahrens, et al. [1984] reports a study of the stability of viscometric flow using the type of short memory introduced by Akbay, Becker, Krozer and Sponagel [1980]. The instability found by Akbay, et al. can be identified as a loss of evolution leading to the catastrophic short wave instability of Hadamard type whenever

$$\frac{\left[\left[\frac{N_1(\kappa)}{\kappa}\right]'\right]^2 \kappa^3}{\tau(\kappa)\tau'(\kappa)} > 16\,, \tag{2.3}$$

where κ is the shear rate, $\tau(\kappa)$ is the shear stress and $N_1(\kappa)$ is the first normal stress. Catastrophic instabilities to short waves of this type may be characteristic for some of the types of instability called "melt fracture". Ahrens, et al. addressed the question of justification for the short memory assumption and find that it cannot be justified for some of the more popular rheological models. The left side of (2.3) reduces to the square of the recoverable shear (N_1^2/τ^2) when the variation of N_1/κ^2 and τ/κ is small. W. Gleissle [1982] found that flow instabilities (melt fracture) commenced in 14 very different type polymer melts and solutions when the recoverable shear varied 4.36 - 5.24 with a mean 4.63. This seems to be in rather astonishing agreement with the criterion (2.3).

Hadamard instabilities may be endemic in the theory of flow of viscoelastic fluids with instantaneous elasticity.

3. Loss of Evolution and Change of Type for Models with Hyperbolic Vorticity.

We could consider all the models which give rise to a vorticity equation which can be hyperbolic in the steady case. These were identified by Joseph, Renardy and Saut [1985] as JRS models. They include all constitutive models whose principal parts are of Oldroyd type

$$\lambda \frac{\tau}{t} = 2\eta\, D[\mathbf{u}] + \boldsymbol{\ell}\ , \tag{3.1}$$

where $\boldsymbol{\ell}$ is of lower order (see JRS) and where λ is the relaxation time, τ is the extra stress, η is the elastic viscosity, $D[\mathbf{u}]$ is the symmetric part of the velocity gradient, and

$$\frac{\tau}{t} = \frac{\partial\tau}{\partial t} + (\mathbf{u}\nabla)\tau + \tau\boldsymbol{\Omega} - \boldsymbol{\Omega}\tau - a(D\tau + \tau D)\ , \tag{3.2}$$

where $\Omega = \frac{1}{2}(\nabla\mathbf{u} - \nabla\mathbf{u}^T)$ is the skew symmetric part of $\nabla\mathbf{u}$ and a [-1,1].

The lower order terms $\boldsymbol{\ell}$ in (3.1) may depend on $\mathbf{u}$ and τ, but not on their derivatives. The upper convected, corotational and lower convected Maxwell models arise when $\boldsymbol{\ell} = -\tau$ and a = [1,0,-1]. Different models can be obtained by different theories of the lower order terms $\boldsymbol{\ell}$. One version of the model by Phan-Thien and Tanner [1977] may be expressed by (3.1) with $\boldsymbol{\ell} = -(1 + c\ \mathrm{tr}\ \tau)\tau$ where c is a constant. The Johnson-Segalman model [1977] with an exponential kernel is a special case of Phan-Thien and Tanner with c = 0 and in fact is one of the Maxwell type of Oldroyd models. A simple Giesekus model [1982] is given by (3.1) with a = 1, $\boldsymbol{\ell} = -(\tau + \frac{\alpha}{\mu}\tau^2)$ where $0 \le \alpha \le 1$ is a constant related to the relative mobility tensor.

We now turn to the analysis of evolutionarity for systems governed by (3.1). The method used is a kind of linearized stability analysis for short waves of the type already given in Section 2.

We shall study the problem of evolution of the ten field variables $[\mathbf{u},\tau,p]$ satisfying

$$\lambda \frac{\tau}{t} = \eta(\nabla\mathbf{u} + \nabla\mathbf{u}^T) + \boldsymbol{\ell}\ ,$$

$$\rho\left[\frac{\partial\mathbf{u}}{\partial t} + (\mathbf{u}\cdot\nabla)\mathbf{u}\right] + \nabla p - \mathrm{div}\ \tau = 0\ , \tag{3.3}$$

$$\mathrm{div}\ \mathbf{u} = 0\ ,$$

where $\frac{\tau}{t}$ is given by (3.2). We decompose the motion

$$[\mathbf{u},\tau,p] = [\hat{\mathbf{u}},\hat{\tau},\hat{p}] + [\mathbf{u}',\tau',p'] ,$$

where the roof functions satisfy (3.3) and the prime functions are small. We take the liberty of calling the roof flow basic; but in fact it is an arbitrary solution of (3.3). After linearizing, we find that

$$\lambda[\frac{\partial\tau'}{\partial t} + (\hat{\mathbf{u}}\cdot\nabla)\tau' + \hat{\tau}\Omega' - \Omega'\hat{\tau} - a(D'\hat{\tau} + \hat{\tau}D')] = 2\eta D' + \ell' ,$$

$$\rho\left[\frac{\partial\mathbf{u}'}{\partial t} + (\hat{\mathbf{u}}\cdot\nabla)\mathbf{u}'\right] + \nabla p' - \operatorname{div}\tau' = -\rho(\mathbf{u}'\cdot\nabla)\hat{\mathbf{u}} , \tag{3.4}$$

$$\operatorname{div}\mathbf{u}' = 0 ,$$

where ℓ' does not involve derivatives of $\mathbf{u}'$, τ'.

We next fix our attention at a point x_0 of the field and define $\chi = x - x_0$. Then we imagine that the basic flow and the derivatives of it in (3.3) are constant and equal to their value at x_0. We may then represent the cartesian components of the disturbance as

$$[\mathbf{u}',\tau',p'] \approx [\omega_i,\sigma_{ij},q]\exp i(k_\ell\chi_\ell - \omega t) ,$$

where the ten amplitudes $[\omega_i,\sigma_{ij},q]$ depend on the basic flow at x_0. The amplitudes are governed by

$$c\sigma_{ij} - \frac{1}{2}\hat{\tau}_{i\ell}[(1-a)\omega_\ell n_j - (1+a)\omega_j n_\ell] - \frac{1}{2}[\omega_i n_\ell(1+a) - \omega_\ell n_i(1-a)]\hat{\tau}_{\ell j}$$

$$+ \mu(n_j\omega_i + n_i\omega_j) = O(1/k) ,$$

$$-\rho c\omega_i = -qn_i + \sigma_{ij}n_j , \tag{3.5}$$

$$\omega_i n_i = 0 ,$$

where $\mathbf{n} = \mathbf{k}/k$, $k = \sqrt{k_1^2 + k_2^2 + k_3^2}$ and

$$c = \frac{\omega}{k} - \hat{\mathbf{u}}\cdot\mathbf{n} . \tag{3.6}$$

These equations were first derived by Rutkevitch [1970] for the three values a = [1,0,-1]. When $k \to \infty$, the problem (3.5) can be regarded as an eigenvalue problem for ω/k or c (see (2.2)).

We can find the ten amplitudes $[\omega_i, \sigma_{ij}, q]$ if and only if the determinant Δ of the coefficients vanishes, where

$$\Delta = \left[-\rho c^2 + \mu - \frac{1}{2}\,\tau_{22}(1-a) + \frac{1}{2}\tau_{11}(1+a)\right]\left[-\rho c^2 + \mu + \frac{1}{2}\tau_{11}(1+a) - \frac{1}{2}\tau_{33}(1-a)\right]. \tag{3.7}$$

The derivation of (3.7) follows along lines set down by Rutkevitch; special coordinates are selected such that $n_1 = 1$, $n_2 = n_3 = 0$, $\hat{\tau}_{23} = 0$ and $\hat{\tau}_{22} > \hat{\tau}_{33}$.

The nontrivial values c^2 are then given by

$$c_+^2 = \frac{1}{\rho}\left[\mu + \frac{1}{2}\,\hat{\tau}_{11}(1 + a) - \frac{1}{2}\hat{\tau}_{33}(1 - a)\right],$$

$$c_-^2 = \frac{1}{\rho}\left[\mu + \frac{1}{2}\hat{\tau}_{11}(1 + a) - \frac{1}{2}\hat{\tau}_{22}(1 - a)\right]. \tag{3.8}$$

Departing slightly now from Rutkevitch, using the result proved in Section 4, we call c the velocity of propagation of wave fronts of short waves of vorticity.

In any case, one has $c^2 = f$, where f is real valued. If $f > 0$, we get propagation. If $f < 0$, then

$$c = \pm i\sqrt{f} = \pm i\,\mathrm{Im}\left[\frac{\omega}{k}\right]. \tag{3.9}$$

Equation (3.9) shows that if $f < 0$, then there exist short waves of rapidly growing amplitude, the flow undergoes a Hadamard instability.

If we now suppose that at x_0 the system is in principal coordinates of $\hat{\tau}$, the eigenvalues of $\hat{\tau}$ satisfying

$$\hat{\tau}_1 \geq \hat{\tau}_2 \geq \hat{\tau}_3 . \tag{3.10}$$

Then $f > 0$, (and the system is of evolution type, stable to short waves) if and only if

$$\mu + \frac{a}{2}(\hat{\tau}_1 + \hat{\tau}_3) - \left[\frac{\hat{\tau}_1 - \hat{\tau}_3}{2}\right] > 0 . \tag{3.11}$$

Among the Maxwell models ($\boldsymbol{\ell} = \boldsymbol{\tau}$) only the upper ($a = 1$) and lower ($a = -1$) convected models are always evolutionary. This follows from the integral form of these two models. The integrals are expressed by positive definite tensors restricting the range of τ to evolutionary regions (see JRS, 1985). Dupret and Marchal [1985] have shown that if the criterion for evolution is satisfied initially it will not fail subsequently. We will show in Section 7 that Maxwell fluids $\boldsymbol{\ell} = -\tau$ and with $a \neq \pm 1$ can lose evolution in certain flows. The models Phan-Thien and Tanner, [1977], Johnson and Segalman [1977], Leonov [1976] and Giesekus [1982] may also lose evolution in certain flows.

4. Evolution of the Vorticity

The vorticity equation for (3.3) in 3-D flows may be written as (see (6.4) in Joseph, [1985]):

$$\rho\left[\frac{\partial^2\zeta_k}{\partial t^2} + 2(\mathbf{u}\cdot\nabla)\frac{\partial\zeta_k}{\partial t} + u_e u_j \frac{\partial^2\zeta_k}{\partial x_e \partial x_j}\right] + \frac{1}{2}(a - 1)e_{kej}\tau_{jq}\frac{\partial[\mathrm{curl}\ \boldsymbol{\zeta}]_q}{\partial x_e}$$

$$- \frac{1}{2}(a + 1)\tau_{mp}\frac{\partial^2\zeta_k}{\partial x_m \partial x_p} - \frac{\eta}{\lambda}\nabla^2\zeta_k + \ell_k = 0\ , \tag{4.1}$$

where e_{kej} is the alternating tensor and ℓ_k are all terms of order lower than two derivatives of the vorticity $\boldsymbol{\zeta} = \mathrm{curl}\ \mathbf{u}$.

The analysis for stability to short waves which was given in Section 3 may be applied to (4.1). We find exactly the same formula $\Delta = 0$ given by (3.7).

It follows that the quasilinear system (3.3) is of evolution type if and only if the vorticity equation (4.1) is of evolution type.

5. First Order Quasilinear Systems for Plane Flow

In plane flow, we have six equations in six unknowns

$$\begin{aligned}
&\sigma_t + u\sigma_x + v\sigma_y + \tau(v_x - u_y) - a[2\sigma u_x + \tau(u_y + v_x)] - 2\mu u_x = \ell_1\ , \\
&\tau_t + u\tau_x + v\tau_y + \frac{1}{2}(\sigma - \gamma)(u_y - v_x) - \frac{a}{2}(\sigma + \gamma)(u_y + v_x) - \mu(u_y + v_x) = \ell_2\ , \\
&\gamma_t + u\gamma_x + v\gamma_y + \tau(u_y - v_x) - a[2\gamma v_y + \tau(u_y + v_x)] - 2\mu v_y = \ell_3\ , \\
&\rho(u_t + uu_x + vu_y) + p_x - \sigma_x - \tau_y = 0\ , \\
&\rho(v_t + uv_x + vv_y) + p_y - \tau_x - \gamma_y = 0\ , \\
&u_x + v_y = 0\ ,
\end{aligned} \tag{5.1}$$

where $\tau = \begin{bmatrix} \sigma & \tau \\ \tau & \gamma \end{bmatrix}$, $\mathbf{u} = (u,v)$, where ℓ_1, ℓ_2, ℓ_3 depend on τ and possibly on $\mathbf{u}$, but not on their derivatives.

The analysis of evolution follows exactly along the lines laid out in Section 3 and Section 6. Since there are only two normal stresses, we replace (3.10) with $\hat{\tau}_1 \geq \hat{\tau}_2$ and (3.11) becomes

$$\mu + \frac{a(\hat{\tau}_1 + \hat{\tau}_2)}{2} - \frac{\hat{\tau}_1 - \hat{\tau}_2}{2} > 0 . \tag{5.2}$$

There is only one speed of propagation (3.8) in the plane. The condition (5.2) for evolution is necessary and sufficient for the stability of solutions of (5.1) to short waves. The same condition, but written relative to general coordinates,

$$\begin{gathered} \tau^2 - \left[\mu - \frac{\gamma}{2}(1 - a) + \frac{\sigma}{2}(1 + a)\right]\left[\mu - \frac{\sigma}{2}(1 - a) + \frac{\gamma}{2}(1 + a)\right] < 0 , \\ \frac{1}{2}\gamma(1 - a) - \frac{1}{2}\sigma(1 + a) - \mu < 0 \end{gathered} \tag{5.3}$$

was derived by Joseph, Renardy and Saut [1985] for the vorticity equation associated with (5.1) (see also Section 6)

$$\begin{aligned} \lambda[\rho \frac{\partial^2\zeta}{\partial t^2} &+ 2\rho(\mathbf{u}\cdot\nabla)\frac{\partial\zeta}{\partial t} + \left[\rho u^2 - \mu - \frac{\sigma}{2}(1 + a) + \frac{\gamma}{2}(1 - a)\right]\frac{\partial^2\zeta}{\partial x^2} \\ &+ 2(\rho uv - \tau)\frac{\partial^2\zeta}{\partial x\partial y} + \left[\rho v^2 - \mu + \frac{\sigma}{2}(1 - a) - \frac{\gamma}{2}(a + 1)\right]\frac{\partial^2\zeta}{\partial y^2}] \\ &= \tilde{\ell}_1 \text{ (of lower order)} . \end{aligned} \tag{5.4}$$

Condition (5.2) (or(5.3)) implies that the unsteady equation for vorticity is hyperbolic.

The loss of evolution of system (5.1) should not be confused with the possible change of type of the steady problem. We shall examine the connection with these two phenomena now. The difference between the steady and unsteady problem is most easily explained in terms of the vorticity. The analysis of steady problems for the quasilinear system (5.1) may be written as

$$\mathbf{H}\mathbf{q}_x + \mathbf{J}\mathbf{q}_y + \boldsymbol{\ell} = 0 .$$

where $\mathbf{q}$ is a column vector with components $[u,v,\sigma,\gamma,\tau,p]$ and $\mathbf{H}$, $\mathbf{J}$, $\boldsymbol{\ell}$ depend on $\mathbf{q}$ but not on its derivatives. To analyze the type of this system we look for characteristics $\theta(x,y) = \text{const.}$, $\theta_x dx + \theta_y dy = 0$.

The analysis is straightforward. The characteristics are given by $\frac{dy}{dx} = \alpha$, where α is a solution of

$$\det[-\alpha H + J] = 0 . \tag{5.5}$$

This leads us to (11.2) of JRS:

$$(1 + \alpha^2)(-\alpha u + v)^2\{\rho(-\alpha u + v)^2 + \frac{(\Upsilon - \sigma)}{2}(\alpha^2 - 1)$$

$$+ 2\tau\alpha - (\alpha^2 + 1)(\mu + a\frac{(\Upsilon + \sigma)}{2})\} = 0 . \tag{5.6}$$

There are imaginary roots $\alpha = \pm i$, double real roots along streamlines $\alpha = \frac{v}{u}$ and two roots for the last factor:

$$\alpha = \frac{B}{A} \pm \frac{\sqrt{B^2 - AC}}{A} , \tag{5.7}$$

where

$$A = \mu - \rho u^2 + \frac{1}{2}\sigma(1 + a) - \frac{1}{2}\Upsilon(1 - a) ,$$

$$B = \tau - \rho uv , \tag{5.8}$$

$$C = \mu - \rho v^2 + \frac{1}{2}\sigma(a - 1) + \frac{1}{2}\Upsilon(a + 1) .$$

Wherever

$$B^2 - AC = -\mu^2 + \rho[\mu + a\sigma + a\Upsilon](u^2 + v^2) + \frac{\rho}{2}(\Upsilon - \sigma)(u^2 - v^2)$$

$$+ \tau^2 + \frac{\sigma^2}{4}(1 - a^2) + \frac{\Upsilon^2}{4}(1 - a^2) - \mu a(\sigma + \Upsilon) - 2\rho\tau uv > 0 ,$$

we have two more real characteristics.

Our system (5.5) is therefore of a mixed type: it has imaginary characteristics and therefore is not hyperbolic; it has real characteristics and therefore is not elliptic. This is not an unusual situation in fluid mechanics (the steady Euler equations of incompressible inviscid fluids are of mixed type) but gives rise to mathematical difficulties: the study of such systems is not well developed (see Saut [1985]). What is new is the fact that the characteristics associated with the last factor in (5.6) can be real or complex in the flow. JRS showed that the roots (5.7) are in fact associated with the steady vorticity equation. This equation is either hyperbolic ($B^2 - AC > 0$) or elliptic ($B^2 - AC < 0$). The roots can be elliptic in one region of flow and hyperbolic in other regions, as in the case of transonic flow. The other characteristics have a simple interpretation: the imaginary roots $\alpha = \pm i$ are associated with the equation $\Delta\psi = -\zeta$, where ζ is the vorticity and ψ the stream function.

6. Vorticity in Plane Flow

In plane flow, there is one nonzero component of vorticity satisfying

$$\rho \frac{\partial^2\zeta}{\partial t^2} + 2\rho(\mathbf{u}\cdot\nabla)\frac{\partial\zeta}{\partial t} - A\frac{\partial^2\zeta}{\partial x^2} - 2B\frac{\partial^2\zeta}{\partial x\partial y} - C\frac{\partial^2\zeta}{\partial y^2} + \ell = 0 , \tag{6.1}$$

where ℓ is of lower order and A, B, C are defined by (5.4), i.e.,

$$A = -\rho u^2 + \mu + \frac{\sigma}{2}(1 + a) - \frac{\gamma}{2}(1 - a) ,$$

$$C = -\rho v^2 + \mu - \frac{\sigma}{2}(1 - a) + \frac{\gamma}{2}(a + 1) , \tag{6.2}$$

$$B = \tau - \rho uv .$$

The analysis of evolution is most easily framed relative to (6.1). Let us start with a general definition.

A linear partial differential operator of second order

$$L\zeta = P(\mathbf{x},t,\zeta_{tt},\zeta_{tx_1},\ldots,\zeta_{tx_n},\zeta_{x_1x_1},\zeta_{x_1x_2},\ldots,\zeta_{x_nx_n}) + \text{lower order terms}$$

is evolutionary with respect to t in some domain D of $\mathbb{R}^n$ if for every unit vector $\mathbf{k} = (k_1,\ldots,k_n)$ in $\mathbb{R}^n$, for any t $\mathbb{R}$ and any $\mathbf{x}$ D, the quadratic polynomial in α

$$P(\mathbf{x},t,-\alpha^2,-\alpha k_1,\ldots,-\alpha k_n,-k_1^2,-k_1k_2,\ldots,-k_n^2) = 0$$

has only <u>real zeros</u>. In the case of constant coefficients, this definition implies that there are no plane wave solutions with arbitrarily large amplitude, i.e., there are no Hadamard instabilities.

The polynomial P = 0 evaluated for (6.1) becomes

$$\rho\alpha^2 + 2\alpha\lambda(uk_1 + vk_2) - Ak_1^2 - 2Bk_1k_2 - Ck_2^2 = 0 .$$

This must have real zeroes for every unit vector $\mathbf{k} = (k_1,k_2)$. This leads to

$$(A + \rho u^2)k_1^2 + 2(B + \rho uv)k_1k_2 + (C + \rho v^2)k_2^2 > 0, \; \forall \; k_1,k_2 ,$$

which implies

$$A + \rho u^2 > 0 \text{ and } (B + \rho uv)^2 - (C + \rho v^2)(A + \rho u^2) < 0 .$$

Using (6.2), this is equivalent to (5.3).

The same relationship arises from analysis of stability to short waves. We fix $\mathbf{u},\boldsymbol{\tau}$ (hence A,B,C) at their values at x_0 and put $\ell = 0$ and introduce $\boldsymbol{\chi} = \mathbf{x} - \mathbf{x}_0$, writing

$$\zeta(x,y,t) = \hat{\zeta}(x_0,y_0)\exp i[k_1(x - x_0) + k_2(y - y_0) - \omega t]. \tag{6.3}$$

We find that, with $k^2 = k_1^2 + k_2^2$

$$c^2 = \left[\mu + (\sigma + \Upsilon)\frac{a}{2}\right]k^2 + \frac{1}{2}(\sigma - \Upsilon)(k_1^2 - k_2^2) + 2\tau k_1k_2 , \tag{6.4}$$

where

$$c^2 = \omega^2 + 2\omega(uk_1 + vk_2) + u^2k_1^2 + v^2k_2^2 + 2uvk_1k_2 .$$

For evolution it is necessary that $c^2 > 0$ for all $k_1,k_2 \quad \mathbb{R}$ when $k = (k_1^2 + k_2^2)^{1/2} \to \infty$. If $c^2 < 0$, then

$$\text{Im}\,\frac{\omega}{k} = \pm \text{ positive constant}$$

and we have a nasty instability to short waves. The criterion $c^2 > 0$ is exactly (5.3). When expressed in principal coordinates $\sigma \geq \Upsilon$ and $\tau = 0$, we find that the vorticity is of evolution type provided that

$$\mu + \frac{a(\sigma + \Upsilon)}{2} + \frac{\Upsilon - \sigma}{2} > 0. \tag{6.5}$$

We may relate the criterion for a change of type in steady flow to the criterion for loss of evolution. Consider $\Delta \underset{\text{def}}{=} B^2 - AC$ when $\rho = 0$. Therefore,

$$\Delta = -f_1 f_2 + \tau^2 ,$$

$$f_1 = \mu + a \frac{\gamma + \sigma}{2} - \frac{\sigma - \gamma}{2} , \qquad (6.6)$$

$$f_2 = \mu + a \frac{\gamma + \sigma}{2} - \frac{\sigma - \gamma}{2} .$$

If we suppose that the system (5.1) (or equivalently the unsteady vorticity equation (6.1)) is of evolution type, then we have (5.3) which clearly implies $B^2 - AC < 0$ for $\rho = 0$, i.e., the steady vorticity equation with $\rho = 0$ is elliptic. Thus hyperbolicity of the steady vorticity equation with $\rho = 0$ implies that the full quasilinear system (5.1) is not evolutionary.

The converse is not true. The equation for vorticity with $\rho = 0$ in the steady case is elliptic when

$$\tau^2 - \left[\mu + \frac{\sigma - \gamma}{2} + a \frac{(\sigma + \gamma)}{2}\right]\left[\mu + \frac{\gamma - \sigma}{2} + a \frac{(\sigma + \gamma)}{2}\right] < 0 .$$

This inequality with $a \neq 0$ does not imply the condition (5.3) for evolution when viewed in principal coordinates $\sigma \geq \gamma$, $\tau = 0$; for example, (5.3) is violated when $a < 0$; $\tau = 0$; $\gamma >> 1$; $\sigma >> 1$; $0 < \sigma - \gamma << 1$. To understand this, consider the equation

$$\frac{\partial^2 \phi}{\partial t^2} - A \frac{\partial^2 \phi}{\partial x^2} - C \frac{\partial^2 \phi}{\partial y^2} = 0 . \qquad (6.7)$$

The steady equation is elliptic when $AC > 0$. But the unsteady equation is evolutionary (hyperbolic) with respect to t, if and only if $AC > 0$ and $A > 0$. If $AC > 0$ and $A < 0$, (6.7) is an elliptic equation!

We recall that the quasilinear system (5.1) is evolutionary if and only if (6.1) is evolutionary. We can study loss of evolution by using results from the study of change of type in steady inertialess flow. It is perhaps useful to remark that we must have a loss of evolution, instability to short waves, whenever the vorticity of an inertialess steady flow becomes hyperbolic. Conversely, if the vorticity of an inertialess steady flow is elliptic and $A > 0$, where $A = \mu + \frac{\sigma - \gamma}{2} + a \frac{\sigma - \gamma}{2}$, then the system (5.1) is evolutionary.

All of the models considered here, except the upper and lower convected Maxwell models may change type in an inertialess steady flow.

7. Examples Taken From Linear Theory

In fact, the theory of evolution is based on equations linearized on an arbitrary flow, called basic. We may evaluate the criteria for evolution and change of type on the basic flow. Many examples of this procedure were given in JRS [1985] and by Yoo, Ahrens and Joseph [1985] for the study of change of type in steady flow.

It is of interest to examine the relationship of change of type in steady flow to the study of short wave instability in unsteady flow. Section 11 of JRS gives analysis for change of type in motions for an upper convected Maxwell model perturbing shear flow, extensional flow, sink flow and circular Couette flow. All these problems are elliptic when $\rho = 0$ and all undergo a change of type for $\rho \neq 0$.

A similar type of analysis, using an upper convected Maxwell model, was given by Yoo, Ahrens and Joseph [1985] for the three dimensional sink flow and by Yoo, Joseph and Ahrens [1985] for Poiseuille flow in a channel with wavy walls. These flows also change type when $\rho \neq 0$ and are always evolutionary.

It is of interest to study these problems in cases in which it is possible to lose evolution. We shall examine the examples treated in JRS for Oldroyd models ($a \neq \pm 1$, $\ell = -\tau$ in (3.1)) and some new examples. The corotational Maxwell model ($a = 0$) seems to lose evolution at the lowest levels of stress (the smallest Weissenberg numbers).

7.1 Simple Shear Flow

(a) Oldroyd models (JRS p. 244). For simple shear flows of Oldroyd models we find that $u = \kappa y$, $v = 0$, $\tau = \eta\kappa/D$, $D = 1 + \kappa^2\lambda^2(1 - a^2)$, $\sigma = \lambda\kappa(a + 1)$, $\gamma = \lambda\kappa(a - 1)$. The steady vorticity equation for the linear perturbation is hyperbolic in a strip outside the origin defined by

$$\rho\kappa^2 y^2 > \frac{\eta}{\lambda} \frac{1 + \lambda^2\kappa^2}{1 + \lambda^2\kappa^2(1 - a^2)} \ . \tag{7.1}$$

When $\rho = 0$ we cannot satisfy this inequality. The steady vorticity equation with $\rho = 0$ is always elliptic. Moreover $A = \mu + \frac{\sigma - \gamma}{2} + a\,\frac{(\sigma + \gamma)}{2} = \mu + \lambda\kappa(1 + a^2) > 0$. From Section 6, the linear systems perturbing shear flows are evolutionary.

(b) A Giesekus model ($a = 1$, $\ell = -(\tau + \frac{\alpha}{\mu}\tau^2)$).

The system (5.1) has the form:

$$
\begin{aligned}
&\sigma_t + u\sigma_x + v\sigma_y - 2\sigma u_x - 2\tau u_y - 2\mu u_x = -\frac{\alpha}{\eta}(\sigma^2 + \tau^2) - \frac{\sigma}{\lambda}\,,\\
&\tau_t + u\tau_x + v\tau_y - \gamma u_y - \sigma v_x - \mu(u_y + v_x) = -\frac{\alpha}{\eta}(\sigma\tau + \gamma\tau) - \frac{\tau}{\lambda}\,,\\
&\gamma_t + u\gamma_x + v\gamma_y - 2\tau v_x - 2\gamma v_y - 2\mu v_y = -\frac{\alpha}{\eta}(\tau^2 + \gamma^2) - \frac{\gamma}{\lambda}\,,\\
&\rho(u_t + uu_x + vu_y) + p_x - \sigma_x - \tau_y = 0\,,\\
&\rho(v_t + uv_x + vv_y) + p_y - \tau_x - \gamma_y = 0\,,\\
&u_x + v_y = 0\,.
\end{aligned} \tag{7.2}
$$

In simple shear flow, we find $u = \kappa y$, $v = 0$, $\tau = \frac{\eta\kappa}{1 + \kappa^2\lambda^2}$, $\sigma = \lambda\kappa\tau$, $\gamma = -\sigma$.

The steady vorticity equation for the linearized flow is hyperbolic in a strip outside the origin defined by

$$\rho\kappa^2 y^2 > \mu\,.$$

This inequality cannot be satisfied for $\rho = 0$ and the steady equation for vorticity is always elliptic in inertialess flows. Moreover, $A = \mu + \sigma = \mu + \frac{\lambda\eta\kappa^2}{1 + \kappa^2\lambda^2} > 0$, and the linear system perturbing shear flow is of evolution type.

7.2 Poiseuille Flow of an Oldroyd Model

In this example, $p_x = -K$, $K > 0$ is a prescribed constant and

$$\tau_y = -K \ ,$$

$$\tau(1 + a)u_y = \sigma/\lambda \ ,$$

$$\tau(1 - a)u_y = -\gamma/\lambda \ , \tag{7.3}$$

$$\left[\frac{1}{2}(\sigma - \gamma) - \frac{a}{2}(\sigma + \gamma) - \frac{\eta}{\lambda}\right]u_y = -\frac{\tau}{\lambda} \ .$$

Putting $u_y = \kappa = \kappa(y)$, we find that

$$\tau = \frac{\eta\kappa}{\lambda^2(1 - a^2)x^2 + 1} \ ,$$

$$\sigma = \frac{\lambda\eta\kappa^2(1 + a)}{\lambda^2(1 - a^2)\kappa^2 + 1} \ , \tag{7.4}$$

$$\gamma = \frac{\lambda\eta\kappa^2(a - 1)}{\lambda^2(1 - a^2)\kappa^2 + 1} \ .$$

Flows perturbing Poiseuille flow of an Oldroyd fluid will become unstable to short waves and lose evolutionarity wherever the steady vorticity equation with ρ = 0 is hyperbolic. This condition may be expressed using (5.8) as

$$B^2 - AC = -\frac{\eta^2}{\lambda^2} + \tau^2 + \frac{(\sigma^2 + \gamma^2)}{4}(1 - a^2) - \frac{\eta}{\lambda}\, a(\sigma + \gamma) > 0 \ . \tag{7.5}$$

In the present case, this condition reads

$$\tau^2\left[1 + \frac{\lambda^2\kappa^2}{2}(1 - a^4)\right] - 2a^2\eta\kappa\tau - \frac{\eta^2}{\lambda^2} > 0 \ , \tag{7.6}$$

where τ is given by (7.4). After a short computation, the left-hand side of (7.6) is evaluated as a positive coefficient times $-\frac{\lambda^4\kappa^4}{2}(1 - a^4) - \lambda^2\kappa^2 - 1$, hence the inequality (7.6) cannot be satisfied for any a (-1,1).

The flows perturbing Poiseuille flow of an Oldroyd model are always evolutionary, it is only necessary to verify that

$$A = \left[\frac{\eta}{\lambda}\right]\left[\frac{1 + 2\lambda^2\kappa^2}{1 + \lambda^2(1 - a^2)\kappa^2}\right] > 0 \ .$$

7.3 Extensional Flow

(a) Oldroyd Models (JRS p. 244). We find that

$$[u,v,\tau,\sigma,\gamma] = \left[sx,-sy,0,\frac{2\eta s}{p},-\frac{2\eta s}{q}\right]$$

where $p = 1 - 2a\lambda s$, $q = 1 + 2a\lambda s$. We take $s \geq 0$ and small enough so that p and q are bounded from above, $s^2 < \frac{1}{4a^2\lambda^2}$.

The steady vorticity of motions perturbing extensional flow is hyperbolic when

$$\rho s^2x^2\left[\frac{\eta}{\lambda} + \frac{2\eta s(2a^2\lambda s - 1)}{pq}\right] + \rho s^2y^2\left[\frac{\eta}{\lambda} + \frac{2\eta s(1 + 2a^2\lambda s)}{pq}\right]$$

$$> \left[\frac{\eta}{\lambda} + \frac{2\eta s(1 + 2a^2\lambda s)}{pq}\right]\left[\frac{\eta}{\lambda} + \frac{2\eta s(2a^2\lambda s - 1)}{pq}\right] . \tag{7.7}$$

If the second factor on the right is positive, the region outside an ellipse is hyperbolic. We put $\rho = 0$ for inertialess flow. Then we get hyperbolicity for the vorticity with $\rho = 0$ when

$$0 > \left[\frac{\eta}{\lambda} + \frac{2\eta s(1 + 2a^2\lambda s)}{(pq)^2}\right]\left[\frac{\eta}{\lambda} - 2\eta s\right]/pq .$$

Therefore we lose evolution when

$$pq + 2\lambda s(2a^2\lambda s - 1) < 0.$$

That is, when $s > \frac{1}{2\lambda}$. This value of s is in the allowed range $s^2 < 1/4a^2\lambda^2$ provided a (-1,1). The linear systems governing flows which perturb extensional flow are unstable to short waves whenever $s > 1/2\lambda$. Steady flows with inertia change type in the manner specified by (5.4) and are evolutionary when $s < \frac{1}{2\lambda}$. (It is easily verified that A with $\rho = 0$ is equal to $\mu + \frac{2\eta s}{pq} + \frac{4a^2\eta\lambda s^2}{pq} > 0$).

(b) A Model of Phan-Thien and Tanner

The analysis given under (a) above applies here also because tr $\tau = 0$ for extensional flow. This model is nonevolutionary when $s > \frac{1}{2\lambda}$.

(c) A Model of Giesekus

The system (7.2) is satisfied by the following extensional flow: $u = sx$, $v = sy$, $\tau = 0$ where σ and γ are given by

$$\frac{\alpha}{\eta}\sigma^2 + \sigma\left(\frac{1}{\lambda} - 2s\right) - 2s\mu = 0 \,,$$

$$\frac{\alpha}{\eta}\gamma^2 + \gamma\left(\frac{1}{\lambda} + 2s\right) + 2s\mu = 0 \,.$$

It follows that the stresses in extensional flow are given by

$$\sigma_{\pm} = \frac{\eta}{\alpha}\left[\frac{2s\lambda - 1}{\lambda} \pm \left[\left[\frac{1 - 2s\lambda}{\lambda}\right]^2 + \frac{8s\alpha}{\lambda}\right]^{1/2}\right],$$

$$\gamma_{\pm} = \frac{\eta}{\alpha}\left[-\frac{2s\lambda + 1}{\lambda} \pm \left[\left[\frac{1 + 2s\lambda}{\lambda}\right]^2 - \frac{8s\alpha}{\lambda}\right]^{1/2}\right].$$

Since $0 \le \alpha \le 1$, $\left[\frac{1 + 2s\lambda}{\lambda}\right]^2 - \frac{8s\alpha}{\lambda}$ is positive and the stresses double valued.

Let us suppose now, following Giesekus [1982, pg. 79], that the configuration tensor $\mathbf{C} = \mathbf{1} + \frac{\tau}{\mu}$ is positive definite. Then $1 + \frac{\sigma}{\mu} > 0$ and $1 + \frac{\gamma}{\mu} > 0$. These inequalities cannot be satisfied for the negative roots of (7.9) when $s < \frac{1}{2\lambda}$. To show this we set

$$\alpha\left[1 + \frac{\sigma_-}{\mu}\right] = \alpha + 2s\lambda - 1 - [(1 - 2s\lambda)^2 + 8s\alpha\lambda]^{1/2} \underset{\text{def}}{=} f(\alpha) \,.$$

One has

$$f(0) = 2(2s\lambda - 1) < 0 \text{ and } f(1) = -1 \,.$$

On the other hand

$$f'(\alpha) = 1 - \frac{4s\lambda}{[(1 - 2s\lambda)^2 + 8s\alpha\lambda]^{1/2}}$$

and

$$f'(0) = \frac{1 - 6s\lambda}{1 - 2s\lambda}, \qquad f'(1) = \frac{1 - 2s\lambda}{1 + 2s\lambda} > 0 .$$

Since $f'(\alpha) = 0$ has one and only one zero, we see that $f(\alpha) < 0$ for α [0,1].

It is easy to verify that $\alpha\left[1 + \frac{\sigma_+}{\mu}\right] > 0$ for α [0,1] and σ_+ is admissible under the supposed restriction on the eigenvalues of the configuration tensor.

Similarly the reader can easily verify that $\alpha\left[1 + \frac{\gamma_-}{\mu}\right] < 0$ for α [0,1], and that $\alpha\left[1 + \frac{\gamma_+}{\mu}\right] > 0$ for α [0,1] provided $s < \frac{1}{4\lambda}$. It follows that under the assumed restrictions, $\sigma = \sigma_+$ and $\gamma = \gamma_+$ for extensional flow.

The steady vorticity of motions perturbing extensional flow is hyperbolic when

$$\rho s^2x^2(\gamma_+ + \mu) + \rho s^2y^2(\sigma_+ + \mu) - (\sigma_+ + \mu)(\gamma_+ + \mu) > 0 . \qquad (7.10)$$

Since $\mu + \gamma_+$, $\mu + \sigma_+$ are positive, (7.10) describes the exterior of an ellipse.

When $\rho = 0$, (7.10) reduces to

$$(\sigma_+ + \mu)(\gamma_+ + \mu) < 0 ,$$

which is impossible for α [0,1] and $0 \le s < \frac{1}{4\lambda}$: the linear system perturbing extensional flow is always evolutionary in the Giesekus model whenever the configuration tensor is positive. It is perhaps necessary to note that unlike the upper and lower convected Maxwell models, the Giesekus model does not restrict the range of stresses to an evolutionary domain; an extra condition, perhaps inconvenient for numerical analysis, has to be imposed.

7.4 Sink Flow in Three Dimensions (Yoo, Ahrens and Joseph [1985])

Maxwell models allow an irrotational solution for flow into a sink:

$$u = -\frac{Q}{r^2},\ v = 0\ ,$$

where u is the radial component of velocity, $Q > 0$ is the sink strength, v is the other component of velocity. The stresses $\begin{bmatrix} \sigma & \tau \\ \tau & \gamma \end{bmatrix}$ in polar spherical coordinates are given by $\tau = 0$,

$$\frac{\sigma}{\mu} = 4e^{r^3/3\lambda Q}\ r^{-4a} \int_r^\infty s^{4a-1}\ e^{-s^3/3\lambda Q}\ ds\ ,$$

$$\frac{\gamma}{\mu} = -2e^{r^2/3\lambda Q}\ r^{2a} \int_r^\infty s^{-2a-1}\ e^{-s^3/3\lambda Q}\ ds\ .$$

The vorticity of all steady axisymmetric flows perturbing this solution is elliptic at large r where the stresses are weak and the steady flow changes from elliptic to hyperbolic as the radius is decreased past certain critical values. The flows for $a = \pm 1$ are always evolutionary. For $a = 0$, Yoo, Ahrens and Joseph found that the vorticity of steady flows perturbing sink flow becomes elliptic again near the origin.

In the elliptic region near the origin we have (see (5.7))

$$0 > B^2 - AC = -\mu^2 + \rho\left[\mu + \frac{1}{2}(\gamma - \sigma)\right]u^2 + \frac{\sigma^2 + \gamma^2}{4}\ .$$

Moreover,

$$\gamma - \sigma = 6\mu\ \ell n\ r < 0 \text{ as } r \to 0\ .$$

Hence, as $r \to 0$,

$$0 > \frac{\rho}{2}\ u^2(\gamma - \sigma) + \frac{\sigma^2 + \gamma^2}{4}$$

and $\gamma - \sigma$ is a large negative number and the criterion (6.4)

$$\mu + \frac{\gamma - \sigma}{2} < 0$$

for the loss of evolution is satisfied in the elliptic region near the origin. This region is obviously hyperbolic for the steady vorticity with $\rho = 0$. Smooth, stable (i.e., evolutionary) sink flows of corotational Maxwell fluids are impossible.

8. How to Compute a Newtonian Viscosity From Stress Relaxation or Sinusoidal Oscillations Even When it is Zero

We have seen that fluids with instantaneous elasticity may undergo Hadamard instabilities to short waves at high levels of stress (high Weissenberg numbers). We already noted in Section 2 that these short wave instabilities may be avoided by introducing various regularizing terms. One effective method for regularization which is also natural for viscoelastic fluids is to add a viscosity term to the constitutive equation (for an example, see Dupret, Marchal and Crochet, 1985). Many popular models of fluids have a Newtonian viscosity. The models of Jeffreys, Oldroyd, Rouse and Zimm and molecular models of solutions with Newtonian solvents lead to Newtonian contributions to the stress. To make this method useful it is necessary that the viscosity used should be appropriate to the fluid under study.

To decide about elasticity and viscosity we could consider ever more dilute solutions of polymer chains of large molecules in solvents which might be thought to be Newtonian. What happens when we reduce the amount of polymer? There are two good ideas which are in collision. The first idea says that there is always a viscosity and some elasticity with an ever greater viscous contribution as the amount of polymer is reduced. On the other hand, we may suppose that liquid is elastic so that $\mu = 0$ and the viscosity η is the area under the graph of the relaxation function. Since η is finite in all liquids, we have $\eta = G(0)\bar{\lambda}$, where $\bar{\lambda}$ is a mean relaxation time. Maxwell's idea is that the limit of extreme dilution is such that the rigidity $G(0)$ tends to infinity and $\bar{\lambda}$ to zero in such a way that their product η is finite. Ultimately, when the polymer is gone, we are left with an elastic liquid with an enormously high rigidity. This idea apparently requires anomalous behavior because $G(0)$ appears to decrease with polymer concentration when the concentration is finite.

The contradiction between the two foregoing ideas and the apparent anomaly can be resolved by replacing the notion of a single mean relaxation time with a distribution of relaxation times. This notion is well grounded in structural theories of liquids in which different times of relaxation correspond to different modes of molecular relaxation. It is convenient again to think of polymers in a solvent, but now we can imagine that the solvent is elastic, but with an enormously high rigidity. In fact many of the so called Newtonian solvents have a rigidity of the order 10^9 Pascals, which is characteristic of glass, independent of variations of the chemical characteristics among the different liquids (for example, see Harrison, 1976). To find this glassy modulus it is necessary to use very ingenious high frequency devices operating in the range 10^9 Hertz and to supercool the liquids to temperatures near the glassy state. In these circumstances the liquid acts like a glassy solid, the molecular configurations cannot follow the rapid oscillations of stress, the liquid cannot flow. For slower processes it is possible for the liquid to flow and if the relaxation is sufficiently fast the liquid will appear to be Newtonian in more normal flows. For practical purposes there is no difference between Newtonian liquids and liquids with rigidities of order 10^9 and mean relaxation times of 10^{-10}seconds or so. In fact it is convenient to regard such liquids as Newtonian, even though $\mu = 0$ and $\tilde{\mu} = \eta$.

The presence of polymers would not allow the liquid to enter the region of viscous relaxation at such early times. Instead much slower relaxation processes associated with the polymers would be induced. The second epoch of relaxation occurs in a neighborhood of very early times $t = t_1$ (or at very high frequencies). A plateau modulus $G(t_1)$ may be defined at $t = t_1$ or for any t in the neighborhood of t_1. The plateau modulus is not so well defined since G(t) is a rapidly varying function in the neighborhood of $t=t_1$.

The relaxation function may be measured on standard cone and plate rheometers, using, for example, stress relaxation after a sudden strain. Examples of such stress relaxation, taken on a Rheometric System 4 rheometer is shown in Figs. 1 and 2. The rise time of this instrument is roughly 0.01 sec and the more rapid part of the stress relaxation cannot be obtained with such devices. The modulus G_C was measured by Joseph, Riccius and Arney (forthcoming) using a wave speed meter. They measure transit times of impulsively generated shear waves into a viscoelastic liquid at rest. A Couette apparatus is used; the outer cylinder is moved impulsively; the time of transit of the shear wave from the outer to inner cylinder is measured. They set up criteria to distinguish between shear waves and diffusion. One criterion is that transit times δt should be reproducible without large standard deviations and such that $d = c\delta t$, where d is gap size, and c, the wave speed is a constant independent of d. In other words, transit times are independent of gap size. Then, using theoretical results for propagation of shear waves into rest, $c = \sqrt{G_C/\rho}$. We could regard G_C as the plateau modulus or the effective rigidity.

It is clear from Figs. 1 and 2 that the rapidly relaxing part of the shear relaxation function, even ignoring the possibility of enormously fast relaxations in times of order 10^{-10} seconds in the glycerin and water solutions associated with glassy states of the two solutions, is missed out on the data of the Rheometrics four. We may also note that the tail end of G(t) is also not accessible on standard rheometers because the transducers do not work when the levels of stress are too low.

Similar limitations of capacity are characteristic for the gap loading devices used for sinusoidal oscillations in standard cone and plate rheometers. The high frequency devices which are used to determine glassy responses of low molecular weight liquids do not work well for polymeric liquids.

To compute a good value for the Newtonian viscosity even when it is zero we need to find a way to put the part of the viscosity which is associated with rapid relaxation into a Newtonian viscosity. For this it suffices to have, say, $G(t)$ for $.01 < t < \hat{t}$. We can get this from any standard rheometer with a stress relaxation capacity. Or, we could use the complex viscosity, computed by standard rheometers, which have upper limits of 100 rad/sec. In addition it is necessary to measure the zero shear or static viscosity $\tilde{\mu}$, given in Figs. 1 and 2 as the area of the box.

The computation procedure, starting from stress relaxation is as follows.

1. We fit a shear relaxation spectrum to the given relaxation function. From the measured values of $G(t)$, $t_0 \leq t \leq t_1$, we get a theoretical function $\hat{G}(t)$, $0 \leq t \leq \infty$. We should make the curve fitting in an honest way such that

$$\hat{\eta} = \int_0^\infty \hat{G}(t)\,dt$$

is as small as it can honestly be. Of course $\hat{G}(0) < G_c$ ($\leq G(0)$) may be much less.

2. Measure $\tilde{\mu}$, the area of the box.

3. $\mu = \tilde{\mu} - \hat{\eta}$ is the required value of the Newtonian viscosity, larger than it probably should be, but honest.

We could use small sinusoidal oscillation data instead of stress relaxation to compute $\hat{G}(t)$.

This paper has been prepared for Amorphous Polymers Workshop, March 5-8, 1985, held at the Institute for Mathematics and its Applications. The work was supported by the U.S. Army Research Office, Math and by the National Science Foundation, Fluid Mechanics. Many of the results given here are taken from previous works with various collaborators, but most especially from a recent [1985] work with Jean Claude Saut.

References

M. Ahrens, D. D. Joseph, M. Renardy and Y. Renardy, Remarks on the stability of viscometric flow, Rheol. Acta **23** (1984), 345-354.

U. Akbay, E. Becker, S. Krozer and S. Sponagel, Instability of slow viscometric flow, Mech. Res. Comm. **7** (1980), 199-200.

H. Aref, Finger, bubble, tendril, spike. Invited General Lecture, Polish Academy of Science XVII Biennial Fluid Dynamics Symposium, Sobieszeivo, Poland, Sept. 1-6, 1985.

R. B. Bird, B. Armstrong, O. Hassager, Dynamics of Polymeric Liquids. Wiley, 1977.

K. V. Brushlinskii, A. I. Morozov, On the evolutionarity of equations of magnetohydrodynamics taking Hall effect into account, PMM **32**, No. 5 (1968), 957-959.

B. D. Coleman and W. Noll, Foundations of linear viscoelasticity, Rev. Mod. Physics **33**, 239-249 (1961). Erratum op. cit. **36**, 1103 (1964).

F. Dupret, J. M. Marchal, Sur le signe des valeurs propres du tenseur des extra-constraints dans un ecoulement de fluide de Maxwell, forthcoming.

F. Dupret, J. M. Marchal, Proceedings of the fourth workshop on numerical methods in non-Newtonian flows, Spa, Belgium, June 3-5, 1985 (to appear in JNNFM).

F. Dupret, J. M. Marchal, and M. J. Crochet, On the consequence of discretization errors in the numerical calculation of viscoelastic flow, J. Non Newtonian Fluid Mech. **18**, 173-186 (1985).

D. Gidaspow, In "Heat Transfer 1974". Japan Soc. of Mech. Engrs. **VII**, (1974), 163.

H. Giesekus, A simple constitutive equation for polymer fluids based on the concept of deformation-dependent tensorial mobility, J. Non Newtonian Fluid Mech. **11** (1982), 69-109.

W. Gleissle, Stresses in polymer melts at the beginning of flow instabilities (melt fracture) in cylindrical capillaries, Rheol. Acta **21** (1982), 484-487.

G. Harrison, The Dynamic Properties of Supercooled Liquids. Academic Press, 1976.

M. W. Johnson, D. Segalman, A model for viscoelastic fluid behavior which allows non-affine deformation, J. Non Newtonian Fluid Mech. **2** (1977), 255-270.

D. D. Joseph, Hyperbolic phenomena in the flow of viscoelastic fluids. Proceedings of the Conference on Viscoelasticity and Rheology, U of WI (1984), edited by J. Nohel, A. Lodge, M. Renardy, Academic Press, (to appear). See also MRC Report 2782.

D. D. Joseph, Historical perspectives on the elasticity of liquids, J. Non-Newtonian Fluid Mech. **19** (1986), 237-249.

D. D. Joseph, M. Renardy, J. C. Saut, Hyperbolicity and change of type in the flow of viscoelastic fluids, Arch. Rational Mech. Anal. **87** (1985), 213-251.

D. D. Joseph and J. C. Saut, Change of type and loss of evolution in the flow of viscoelastic fluids, J. Non-Newtonian Fluid Mech. **20** (1986), 117-141.

D. D. Joseph, O. Riccius, M. Arney, Wavespeeds and elastic moduli for different liquids, Part 2. Experiments, J. Fluid Mech. **171** (1986), 309-338.

A. G. Kulikovskii, S. A. Regirer, On stability and evolution of the electric current distribution in a medium with nonlinear conductivity, PMM **32**, No. 4 (1968), 76-79.

A. G. Kulikovskii, Surfaces of discontinuity separating two perfect media of different properties. Recombination waves in magnetohydrodynamics, PMM **32**, No. 6 (1968), 1125-1131.

A. I. Leonov, Nonequilibrium thermodynamics and rheology of viscoelastic polymer media, Rheol. Acta **15** (1976), 85-98.

E. H. Lieb, Ph. D. Thesis, University of Delaware, 1975.

R. Lyczkowski, D. Gidaspow, D. Solbrig, C. Hughes, Characteristics and stability analysis of transient one-dimensional two-phase flow equations and their finite difference approximations, Nucl. Sci. Eng. **66** (1978), 378.

N. Phan-Thien, R. I. Tanner, A new constitutive equation derived from network theory, J. Non Newtonian Fluid Mech. **2** (1977), 353-365.

M. Renardy, Singularly perturbed hyperbolic evolution problems with infinite delay and an application to polymer rheology, SIAM J. Math. Anal. **15** (1984), 333-349.

M. Renardy, "A local existence and uniqueness theorem for K-BKZ fluid, Arch. Rational Mech. Anal. **88** (1985), 83-94.

I. M. Rutkevitch, Some general properties of the equations of viscoelastic incompressible fluid dynamics, PMM **33**, No. 1 (1969), 42-51.

I. M. Rutkevitch, The propagation of small perturbations in a viscoelastic fluid, J. Appl. Math. Mech. **34** (1970), 35-50.

P. Saffman, G. I. Taylor, The penetration of a fluid into a porous medium or Heleshaw cell containing a more viscous liquid, Proc. Roy. Soc. **A245** (1958), 312-29.

J. C. Saut, Mathematical problems associated with equations of mixed type for the flow of viscoelastic fluids. Proceedings of the fourth workshop on numerical methods in non-Newtonian flows, Spa, Belgium, June 3-5, 1985 (to appear in JNNFM).

J. C. Saut, D. D. Joseph, Fading memory, Arch. Rational Mech. Anal. **81**, 53-95 (1983).

J. S. Ultman, M. M. Denn, Anomalous heat transfer and a wave phenomenon in dilute polymer solutions, Trans. Soc. Rheology **14**, 307-317 (1970).

J. Y. Yoo, M. Ahrens, D. D. Joseph, Hyperbolicity and change of type in sink flow, J. Fluid Mech. **153** (1985), 203-214.

J. Y. Yoo, D. D. Joseph, Hyperbolicity and change of type in the flow of viscoelastic fluids through channels, J. Non Newtonian Fluid Mech. (1985), to appear.

RUBBERY LIQUIDS IN THEORY AND EXPERIMENT

E.A. Kearsley

Rheology Research
P.O. Box 2101
Montgomery Village Station
Gaithersburg, MD 20879

The word "rubbery" is an adjective widely used and understood to imply mechanical properties similar in a certain way to that of everyday rubber such as that of ordinary rubber bands. "Rubbery" is certainly not equivalent to "elastic" since quartz or steel are highly elastic but are not rubbery. Examples of materials exhibiting the property of "rubberyness" are to be found among bulk amorphous polymers such as polyisobutylene or plasticized polyvinylchloride, and liquids such as common rubber cement. Though it is difficult to say precisely what is meant by the term, most people would agree that it suggests the capacity for extensive stretching and examples of rubbery materials that come to mind are usually polymeric substances as in the list above, that is, they have in common that their microscopic structure includes assemblages of very long, flexible chain molecules. One is led to seek an understanding of the behavior of these materials through an examination of the archetypical rubbery material, vulcanized natural rubber.

Ordinary rubber as one usually encounters it has been vulcanized. Frequently it has also been mixed with other substances such as carbon black and from a chemists point of view it is a rather complex material. Relatively pure natural rubber which has been lightly vulcanized is particularly "rubbery", and in some cases it can be stretched to five or even ten times its original length and yet will recover very nearly to its original length upon removal of the stretching forces. It has long been known that a strip of rubber heats up upon sudden adiabatic stretching. Furthermore, if the strip is stretched isothermally by a fixed weight and then heated externally, the stretch will contract as the temperature of the strip rises. These facts are related through thermodynamic considerations as was shown by Lord Kelvin who pointed out, further, that they imply a tensile force of stretching arising from the "motion of the constituent particles" of the rubber. We can express the main features of the "rubberyness" of vulcanized natural rubber in equilibrium configurations for isothermal or adiabatic deformations from a stress-free state through conventional thermodynamics based on an elastic potential or strain energy function. Such a scheme would appear as follows:

$$\sigma_{ij} = 2\rho x_{i\alpha} x_{j\beta} \frac{\partial W}{\partial C_{\alpha\beta}} ,$$

$$x_i = g_i(X_\alpha),\ F_{i\alpha} = x_{i\alpha} = \frac{\partial g_i}{\partial X_\alpha},\quad C_{\alpha\beta} = x_{i\alpha} x_{i\beta}, \tag{1}$$

$$W = \varepsilon(F,S) \qquad \text{Adiabatic case}$$

$$W = \varepsilon - T\,S \qquad \text{Isothermal case}$$

In these equations, σ is the Cauchy stress, ρ is the density, X_α is a Cartesian coordinate in the (undeformed) stress-free configuration, x_i is a Cartesian coordinate in the deformed state, F is the deformation gradient, C is a Cauchy-Green strain tensor and W plays the mechanical role of a strain energy function. S is the specific entropy and ε is the specific internal energy. From a thermodynamic point of view, for adiabatic deformations W is just the specific internal energy while for isothermal deformations W is a Helmholtz free energy [11, pages 301 and 309].

For the most part, we will restrict our attention in what follows to isothermal deformations of isotropic, incompressible materials. Properly, materials are not strictly incompressible and there should of course be a contribution to the internal energy which depends upon specific volume and leads to an internal pressure just as in ordinary liquids. This part of the internal energy largely determines the bulk modulus of the rubber which is generally several orders of magnitude greater than the shear modulus. For the stretching and shearing deformations we shall be considering, the changes in volume will be ignorably small and the effects of this part of the internal energy may be represented by the arbitrary isotropic pressure of the theory of incompressible elastic media. As a consequence, in this case the elastic equations can be put in the following convenient form:

$$\sigma = -p1 + 2W_1B - 2W_2B^{-1},$$

$$W(I_1,I_2), \quad W_1 = \frac{\partial W}{\partial I_1}, \quad W_2 = \frac{\partial W}{\partial I_2},$$

$$I_1 = trC = tr\, B = \lambda^2 + \mu^2 + \nu^2 \tag{2}$$

$$I_2 = \frac{1}{2}[(tr\, C)^2 - tr\, C^2] = tr\, C^{-1} = tr\, B^{-1} = \frac{1}{\lambda^2} + \frac{1}{\mu^2} + \frac{1}{\nu^2},$$

$$I_3 = \lambda^2\mu^2\nu^2 = 1,$$

$$B = FF^T.$$

In these equations I_1, I_2, I_3, are the conventional scalar invariants of the deformation, B is the left Cauchy Green tensor, W is now the free energy per unit volume and tr represents the trace operator. The principal stretches are represented by λ, μ, and ν.

The actual form of the strain energy function of equation (2) is determined by details of the microscopic structure of the rubber. On the microscopic level we picture vulcanized rubber as a tangled network of very long flexible chains of isoprene units joined together by cross-links, more or less permanent connections which were formed by the vulcanizing treatment and are distributed between or within chains. The "rubberyness" of natural rubber is attributed to the entropic effects of stretching the segments of chain between these crosslinks. The positive stress-temperature coefficient explained by Kelvin through phenomenological thermodynamics can also be appreciated on this microscopic level. Just as, on a macroscopic scale, the tension of a vibrating violin string increases to accomodate an increase of kinetic energy stored in the vibrating string as a result of an increase of amplitude of the vibrations, on a microscopic scale in the stretched rubber, an increase of temperature causes greater thermal fluctuations of the polymer chain segments between crosslinks and a greater tension is needed to keep the segments stretched. If the additional tension is not supplied, the rubber contracts accordingly. With this picture in mind Kelvin's reference to the "motion of the constituent particles" becomes clear.

Structures of varying complexity descriptive of the microscopic networks of rubber can be imagined. Perhaps the simplest statistical model of entropic elasticity results from taking as the contribution to entropy of a segment of molecular chain between crosslinks the mathematical expression for the entropy of a single flexible chain constrained only by the fixed distance between its ends [3, page 56]. This simple model gives a strain energy (Helmholtz free energy) for the isothermal deformation of an incompressible rubber as follows:

$$W = \frac{1}{2} nkTI_1 = \frac{1}{2} nkT(\lambda^2 + \mu^2 + \nu^2), \tag{3}$$

where n is the number of crosslinks per unit volume of rubber, k is Boltzmann's constant, T is absolute temperature. The geometrical form of this strain energy is often called neo-Hookean because it is a natural extension of Hooke's law of infinitesimal elasticity to the case of finite deformations. The stress deformation law for the stress differences between principal directions is given by

$$\sigma_\lambda - \sigma_\mu = nkT(\lambda^2 - \mu^2). \tag{4}$$

It has been known since the late 1940's that this simple model of rubber elasticity is not good enough to describe accurately most actual experimental measurements. Figure 1 shows some results of Treloar which illustrate this point [3, page 118]. The experiment consisted of stretching a sheet of sulfur-vulcanized natural rubber by fixed distributed forces in two orthogonal directions in the plane of the sheet. In the third (thickness) direction perpendicular to the plane of the sheet the force acting is taken as zero. The strain energy which led to equation (4) would require that all the points of the graph fall on a single line passing through the origin. It is clear that the data do not support this. A simple empirical "fix" of this failure of the model can be achieved by replacing the squares of the principal stretches in equation (4) by an arbitrary function of the stretches; thus,

$$\sigma_\lambda - \sigma_\mu = nkT[g(\lambda) - g(\mu)],$$

$$g(\lambda) = \lambda \frac{\partial w(\lambda)}{\partial \lambda}, \qquad (5)$$

$$W = w(\lambda) + w(\mu) + w(\nu),$$

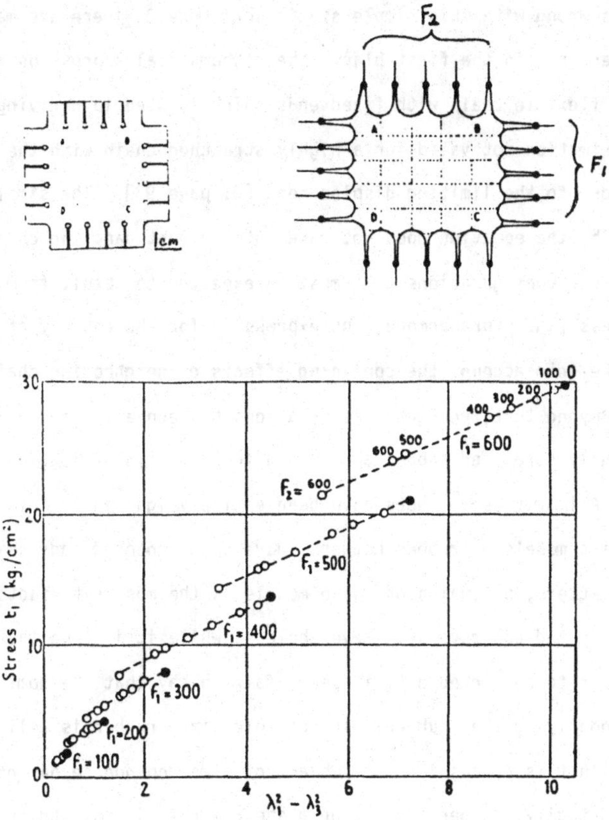

Figure 1

Data of biaxial stretching of a sheet of rubber under dead loading [3]. The points do not fall on a single straight line as they would for a neo-Hookean material. For equal loads of 600 grams the two points are a curiosity implying that a symmetric loading produces an asymmetric stretch [26].

where g(·) is the arbitrary function of stretch. This implies that the strain energy function can be formed by summing a function of stretch, w(·), over the principal stretches of the deformation as in equation $(5)_2$. Strain energies of this type have come to be known as Valanis-Landel forms [14]. Most of the many proposed versions of statistical theories of rubber elasticity lead to some particular Valanis-Landel form of strain energy.

What is wrong with this simple statistical model? There are many possible sources of error. In the first place, the mathematical expression for the entropy of a single flexible chain with fixed ends which is used in deriving equation (4) is an approximation not valid for a highly stretched chain with the separation of its ends close to the limiting displacement [3, page 93]. The simple neat network represented by the equation does not take into account dangling chain ends and loops and other configurations that must be expected to result from the vulcanizing process [1]. Furthermore, the expression for the entropy of a single chain does not take into account the confining effects of neighboring chains in a network [6]. Beyond that, energetic associations between aligned chains are known to occur in highly stretched rubber and even a form of crystallization may occur [3, page 152]. All of these effects have been entirely ignored in this simple model. More elaborate models of rubber networks have been proposed addressing these and many other matters, but which of these models is the most satisfactory remains a topic of study and controversy. Nevertheless, while it is true that important details remain to be worked out, it seems fair to say that the dominant physical mechanism underlying the "rubberyness" of vulcanized rubber is well established.

So far in this discussion the subject of time dependence has not been broached. Actually, rubber supporting a shear stress is not strictly in equilibrium. If a block of rubber is taken from a stress-free, equilibrium configuration to some deformed configuration and held, the stresses required to hold it will decrease with time. The logarithmic plot of stress or of modulus (some ratio of stress to strain) versus time has a certain characteristic form. It is commonly known as a "relaxation curve". Figure 2 is a sketch showing what such a plot looks like. Notice that this graph covers many decades of time and modulus.

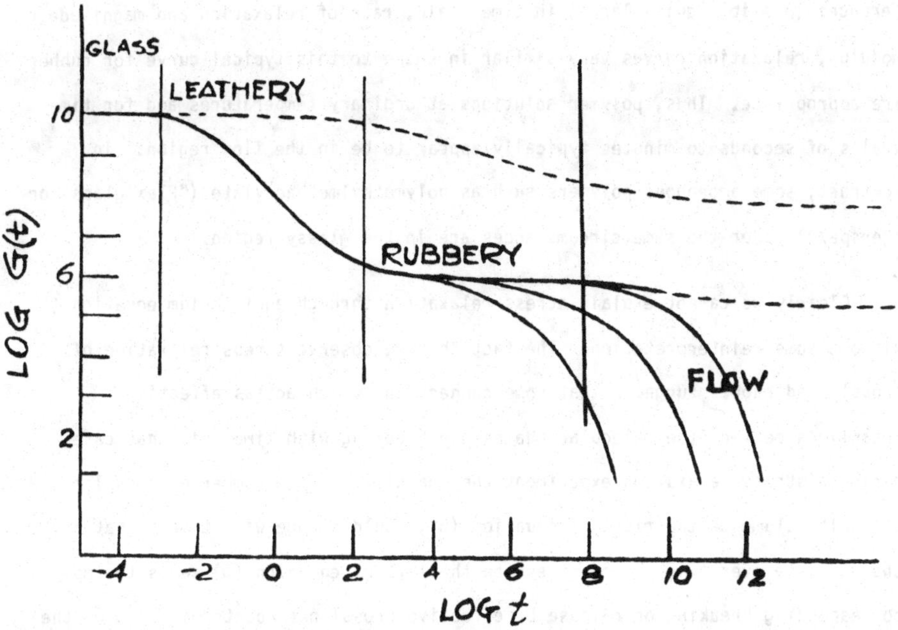

Figure 2

Sketch of a stress relaxation curve for typical rubbers. At the "rubbery plateau" the shear modulus remains relatively constant for the time interval from minutes to hours. The curves in the "flow region" which follows are shown for unvulcanized rubbers and by a dotted line for vulcanized rubber. The relaxation of vulcanized rubber in the flow region is typically about 10% per decade of time. The dotted line at high modulus is for ebonite, a very highly crosslinked crystalline form of rubber.

At very short times there is a plateau of high modulus often called the glassy plateau. With increasing time the modulus decreases and forms a plateau again at a value about three or four orders lower than the glassy plateau covering a range of times from minutes to hours or days. This region is usually called the rubbery plateau and its level and time span are determined by (among other parameters) the density of crosslinks. For the times beyond the rubbery plateau the modulus again falls, typically by about 10% per decade, in what is called the flow region.
For uncrosslinked amorphous polymers and even polymer solutions, allowing for dif-

ferences (possibly quite large) in time scale, rate of relaxation and magnitude of modulus, relaxation curves very similar in shape to this typical curve for rubber are appropriate. Thus, polymer solutions at ordinary temperatures and for time scales of seconds to minutes typically appear to be in the flow region. In contrast, some amorphous polymers such as polymethylmethacrylate ("Plexiglass" or "Perspex") under the same circumstances are in the glassy region.

Clearly we cannot explain stress relaxation through equilibrium equation (5) without some reinterpretation. The fact that we observe stress relaxation of crosslinked rubber suggests that some connections which act as effective crosslinks between the molecular chains are breaking with time. In that case, during a stress relaxation experiment the quantity n, the number of crosslinks per unit volume on the right of equation (5), would change with time. That is, the decrease over time of the stress on the left of equation (5) seems to imply a corresponding breaking or release of effective crosslinks counted in n. On the other hand, to the extent that the rubber will recover upon release of the stress and that after recovery the stress relaxation can be repeated, there must also be a process of forming of effective crosslinks which returns the rubber to the original state upon recovery. Even uncrosslinked materials, insofar as they also are rubbery and behave in this way, seem to encompass mechanisms for forming and breaking effective or "virtual" crosslinks. Early models based on this idea are to be found in the writings of Green and Tobolsky [2], Yamamoto [4] and of Lodge [5].

It is instructive to construct a rudimentary model of the breaking and forming of effective crosslinks and to examine the associated stress relaxation phenomena. To that end let us assume that effective crosslinks (henceforth called simply "bonds") in a rubbery material break at any instant at a rate proportional to the number extant. Let us further assume that there are a finite number of possible sites per unit volume, N, at which bonds can occur and that this number is a constant. Let the rate of forming of bonds be proportional at any instant to the number of these possible sites still unoccupied by crosslinks. The net rate of forming of bonds in a strain free material is then given by

$$\dot{n} = \beta(N - n) - \alpha_0 n, \qquad (6)$$

where $\dot{n}$ is the net rate and n is the number of bonds at any instant, β is the rate of forming bonds, and α_0 is the rate of breaking of the unstrained bonds. Since in equilibrium the net rate must be zero, the number of bonds remains constant and is given by

$$n = n_0 = \frac{N\beta}{\alpha_0 + \beta}, \qquad (7)$$

where we use the symbol n_0 for the number of bonds in the unstrained rubber at equilibrium.

Suppose that at instant t_0 a strain in the material, say a stretch of magnitude λ, is suddenly produced and held. We must now distinguish between the unstrained bonds, call them m, formed after the strain is applied to the material and the strained bonds, n, formed in the original, equilibrium state. We expect only unstrained bonds to be formed and therefore the rate equations at some instant t after the strain is applied are as follows:

$$t > t_0 \qquad \begin{aligned} \dot{n} &= -\alpha_1 n, && \text{strained bonds} \\ \dot{m} &= \beta(N - n - m) - \alpha_0 m, && \text{unstrained bonds} \end{aligned} \qquad (8)$$

where α_1 is the rate of breaking of the strained bonds. These equations can easily be integrated to fit appropriate conditions at instant $t = t_0$; that m shall be zero and n shall be n_0:

$$t > t_0 \qquad \begin{aligned} n &= n_0 e^{-\alpha_1(t-t_0)}, \\ m &= n_0[1 - \frac{\beta}{\alpha_0-\alpha_1+\beta} e^{-\alpha_1(t-t_0)} - \frac{\alpha_0-\alpha_1}{\alpha_0-\alpha_1+\beta} e^{(-\alpha_0+\beta)t-t_0)}]. \end{aligned} \qquad (9)$$

According to our model of the entropy elasticity of chain networks the unstrained bonds do not contribute to the stress. On using equation (5) applied to the stress relaxation due to stretching of a strip of rubber we find that the stress caused by stretching the strip is given by

$$\sigma = nkT\ h(\lambda),$$
$$h(\lambda) = g(\lambda) - g\left(\frac{1}{\sqrt{\lambda}}\right), \tag{10}$$

where n is given by equation (9) and $g(\cdot)$ is the arbitrary function of stretch proposed in equation (5). Equation (10) determines the stress relaxation curves appropriate to this rudimentary model.

The stress relaxation of this crude model is a simple product of an exponential function of time and a nonlinear strain-dependent factor. With experimental data of the proper form one might be able to match stress relaxation curves of a rubbery material to exponential curves of fixed exponent and then, through equations (9) and (10), determine the rates of breaking of bonds as a function of strain. However, since the stress does not reflect the number of unstrained bonds, stress relaxation does not offer a direct measure of the rate of formation of bonds. Notice also that for this simple model the stress, given enough time, will relax exponentially to zero, a characteristic that one might reasonably consider "fluid behavior".

Consider next a double step stress relaxation experiment, that is, the strip originally in stress-free equilibrium is stretched an amount λ at instant t_0 and further stretched to a total stretch μ at instant t_1 (see figure 3). At times after the second stretch there are three types of bonds to be distinguished. Let p denote the number of unstrained bonds, let m denote the number of bonds formed in the time interval between the two stretchings and thus stretched an amount μ/λ and let n denote the number of bonds formed before the initial stretch and thus strained an amount μ. The rate equations for instant t after the second stretching are, accordingly:

$$\dot{n} = -\alpha_2 n,\ \dot{m} = -\alpha_3 m,\ \dot{p} = \beta(N - n - m - p) - \alpha_0 p, \tag{11}$$

where α_2 is the rate of breaking of bonds strained an amount μ and α_3 is the rate for bonds strained an amount μ/λ. The solutions of these rate equations which match with equations (9) at $t = t_1$, the instant of the second step, are

$$n = n_o e^{-\alpha_1(t_1-t_o)} e^{-\alpha_2(t-t_1)}$$
$$m = n_o\left[1 - \frac{\beta}{\alpha_o-\alpha_1+\beta} e^{-\alpha_1(t_1-t_o)} - \frac{\alpha_o-\alpha_1}{\alpha_o-\alpha_1+\beta} e^{-(\alpha_o+\beta)(t_1-t_o)}\right] e^{-\alpha_3(t-t_1)}. \qquad (12)$$

We do not bother to write down the somewhat lengthy expression for p since the unstrained bonds do not contribute to the stress.

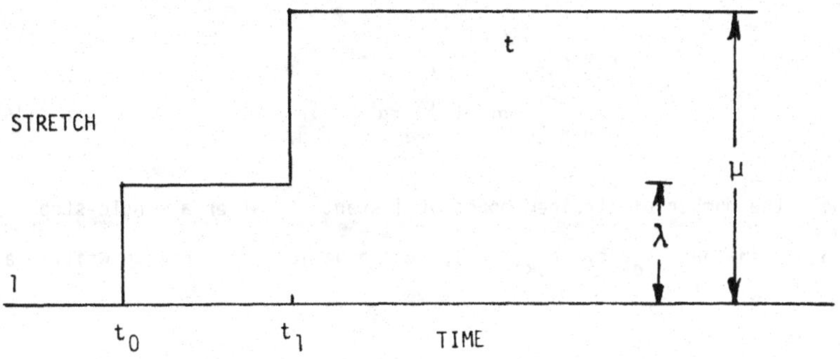

Figure 3

Schematic of the stretch history of a double step stress relaxation experiment. The first stretch λ is followed by a second stretch μ/λ for a total final stretch of μ.

From equations (12), (10a) and (5) we can write down the stress at some instant t after the second step, as follows:

$$\sigma = nkT\, h(\mu) + mkT\, h(\mu/\lambda), \qquad (13)$$

where n and m are the instantaneous numbers of bonds strained by amounts μ and μ/λ, respectively, and are given by equations (12).

We notice immediately that for this model, simple as it is, the stress relaxation after a double step generally cannot be anticipated from the results of single-step stress relaxation. That is to say, in the case of stress relaxation after a double step the rate of forming of bonds plays an important role and appears within the exponents of the expression for stress while in the case of single-step stress relaxation it does not.

In the event that the rate of bond breaking is independent of the strain on the bonds this picture becomes much simpler and neater. In that case we need no longer distinguish the various α's though we must, of course, keep track of the number of bonds under each amount of strain. If we denote the universal rate of breaking of bonds as α, equations (12) can be rewritten in simpler form as follows:

$$n = n_o e^{-\alpha(t-t_o)},$$
$$m = n_o[e^{-\alpha(t-t_1)} - e^{-\alpha(t-t_o)}]. \qquad (14)$$

If we denote the number of strained bonds at instant t after a single-step deformation at instant t_o by $n_{BR}(t,t_o)$, then equations (14) can be written as follows:

$$n = n_{BR}(t,t_o), \quad m = n_{BR}(t,t_1) - n_{BR}(t,t_o), \qquad (15)$$

and the stress relaxation of a double step can be written in terms of a "superposition" of single-step bond relaxation functions

$$\sigma(t) = kT[n_{BR}(t,t_o)h(\mu) + (n_{BR}(t,t_1) - n_{BR}(t,t_o))h(\mu/\lambda)]. \qquad (16)$$

It is a trivial task to interpret this expression as a superposition of three single-step stress relaxation functions.

This superposition principle has a broader applicability. Let us look at the double-step stress relaxation of a more general material for which the stress at any instant is given by a superposition of contributions from all past configurations. To express the instantaneous stress produced by an arbitrary history of extensional deformation for such a material we can write

$$\sigma(t) = -\int_{-\infty}^{t} H_*(\lambda, t-\tau)d\tau, \qquad (17)$$

where $\sigma(t)$ is the stress at some instant t, τ is an instant in the past, $\lambda(t,\tau)$ is the stretch in going from the configuration at instant τ to the configuration at instant t and $-H_*(\lambda,t-\tau)d\tau$ is the contribution of a stretch of amount λ from a configuration at instant τ in the past (the meaning of H and the reason for the minus sign and asterisk subscript will soon become clear). We assume that for a stretch of unity, H_* is zero, that is, we assume that there is no contribution when the material is not actually stretched. If we apply equation (17) to single-step stress relaxation we find that the integrand is zero for values of τ greater than t_1, the instant of stretching, and we get the equation

$$\sigma_{SR}(\lambda,t-t_1) = -\int_{-\infty}^{t_1} H_*(\lambda,t-\tau)d\tau, \tag{18}$$

where $\sigma_{SR}(\lambda,t-t_1)$ is the single-step stress relaxation function for a stretch of λ applied at instant t_1. If we assume that contributions from configurations become negligible as they recede into the past it is easy to see from equation (18) that

$$H_*(\lambda,t-t_1) = \frac{\partial\, \sigma_{SR}(\lambda,t-t_1)}{\partial t}, \tag{19}$$

and we can write

$$H(\lambda,t-t_1) \equiv \sigma_{SR}(\lambda,t-t_1) \tag{20}$$

and the physical significance of H and H_* is made clear.

If we now consider a double-step stress relaxation in which a stretch of λ_1 is applied at instant t_1 followed by a further stretch of λ_2 at instant t_2, we find that the stress at instant t after the second step is given by

$$\sigma(t) = \sigma_{SR}(\lambda_1\lambda_2,t-t_1) + \sigma_{SR}(\lambda_2,t-t_2) - \sigma_{SR}(\lambda_2,t-t_1). \tag{21}$$

The superposition of equation (16) can be recognized in this equation upon substitution of an appropriate special form for H, viz.:

$$H(\lambda,t-t_1) = \sigma_{SR}(\lambda,t-t_1) = kT\, n_{BR}(t,t_1)h(\lambda). \tag{22}$$

What relationship does this superposition have with the classical Boltzmann principle? In linear viscoelasticity, the modulus is a relaxation function independent of amplitude of deformation, that is,

$$\sigma_{SR}(\lambda,t) = 3(\lambda - 1)G(t) \tag{23}$$

where $G(t)$ is the infinitesimal shear modulus. When λ_1 and λ_2 are both close to unity (infinitesimal stretches), we may use the approximation

$$\lambda_1\lambda_2^{-1} \cong \lambda_1 + \lambda_2 - 2 \tag{24}$$

to obtain from equations (21) and (23)

$$\sigma(t) = 3(\lambda_1 - 1)G(t - t_1) + 3(\lambda_2 - 1)G(t - t_2). \tag{25}$$

Equation (25) will be recognized as the classical Boltzmann superposition solution for extensional double-step stress relaxation and it is only valid for deformations within the range of linear mechanical behavior.

The idealized material of equations (1) exhibits no stress relaxation for isothermal deformations. In that material the stress is uniquely a function of the strain, however, it can be generalized in a fairly easy way to a material which does indeed exhibit stess relaxation of a form consistent with equation (21). This can be accomplished by simply taking the stress to be a functional of strain history as in equation (17). There is some insight to be gained, however, by choosing to work instead through a generalizing of equations (1). If in equations (1) we consider the case when the specific internal energy can be written as a function of specific volume and absolute temperature, that is to say, isothermal shearing of the material affects the internal energy only through changes in entropy, then we can write for the stress at any isothermal deformation

$$\begin{aligned} W &= \varepsilon(\upsilon,T) - TS \\ \sigma_{ij} &= g(\upsilon,T)\,\delta_{ij} - x_{i\alpha}x_{j\beta}\rho T\, S_{\alpha\beta} \\ S_{\alpha\beta} &= \frac{\partial S}{\partial C_{\alpha\beta}} \end{aligned} \tag{26}$$

where υ is the specific volume (the inverse of density), S is the specific entropy and C is the Cauchy-Green strain tensor defined in equation (1). The first term in the expression for stress is a "gaslike" term which arises from the change in internal energy with change in volume and which contributes no shear stress. The second term is an "entropic" term and is the sole source of shear stress. For fixed temperature and deformation, nothing in these equations changes with time and no relaxation is possible. Indeed, classical equilibrium thermodynamics does not shed much light on non-equilibrium phenomena and is not suitable for describing stress relaxation.

To introduce the possibility of stress relaxation into this scheme, let us modify the equations (25) by including an addition to the entropy (we shall call it "entaxy") which changes with time when temperature and deformation is fixed [9]. This entaxy will be a superposition of contributions from deformations of past configurations, contributions which will decrease with the passage of time. The passage of time for this purpose will be measured by a "material clock" running at a rate determined by the instantaneous temperature. In summary, the entaxy, ς, is introduced by the following equations and conditions:

$$\begin{aligned} \varsigma(t) &= \int_{-\infty}^{t} S[C(t,\tau), \int_{\tau}^{t} b_T(\xi)d\xi]b_T(\tau)d\tau, \\ S(C(t,\tau),\xi) &> 0, \qquad\qquad (27) \\ \frac{\partial S(C,t)}{\partial t} &< 0 \qquad \text{(fading memory)} \end{aligned}$$

where $S(C,\Delta t)$ is the incremental entaxy due to the deformation C from a configuration at a time interval Δt in the past, $C(t,\tau)$ is the left relative Cauchy-Green tensor at instant t calculated from the past configuration at instant τ as the reference and G_T is the rule of the material clock.

Upon modifying equation (26) by the inclusion of entaxy we find for the stress

$$\begin{aligned} \sigma_{ij}(t) = g(\upsilon,t)\delta_{ij} &+ \qquad\qquad (28) \\ \rho T \int x_{i\alpha}(t,\tau)x_{j\beta}(t,\tau)S_{\alpha\beta}&[C(t,\tau),\int_{\tau}^{t} b_T(\xi)d\xi]b_T(\tau)d\tau, \end{aligned}$$

in exact analogy with the classical equilibrium thermodynamical equations except that an integral over the history of a derivative of the incremental entaxy replaces the derivative of the entropy. This model of a non-equilibrium thermodynamics has the virtue that, whenever state variables are held fixed until all relaxations have ceased, the classical equilibrium thermodynamics is exactly recovered. This model was originally named the "perfect elastic fluid" but it has come to be called more commonly the "BKZ" (sometimes "K-BKZ") model [7,8].

When the restrictions to isothermal deformations and incompressible isotropic materials are made, these equations can be put into a particularly convenient form

$$\sigma_{ij}(t) = -p\,\delta_{ij} + 2\int_{-\infty}^{t} \left[\frac{\partial U}{\partial I_1} C_{ij}(t,\tau) - \frac{\partial U}{\partial I_2} C^{-1}_{ij}(t,\tau)\right] d\tau$$
$$U = U[I_1(t,\tau),\ I_2(t,\tau),\ t-\tau], \qquad (29)$$

where I_1 and I_2 are scalar invariants of the left relative Cauchy-Green strain tensor, C, and are given by the principal stretches as in equation (2).

In light of equation (2), equation (29) may be interpreted as giving the stress as a sum of elastic stresses for deformations to the present configuration of all past configurations, each calculated from a relaxing strain energy. As a result, the stress history for any deformation history can be calculated from single-step stress relaxation data.

Figure 4 is an example of a set of single-step stress relaxation data taken in extension on plasticized polyvinyl chloride plotted as a modulus versus the logarithm of time [8]. Observe that the curves, each of which corresponds to a different extension ratio, do not have a common shape. Some of these curves actually cross over others demonstrating clearly that for this material the single step stress relaxation can not be expressed simply through a product of a function of deformation and a function of time.

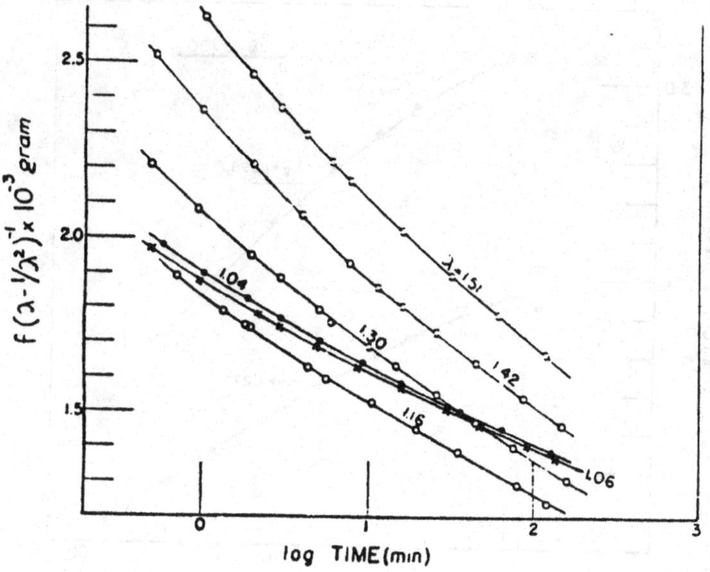

Figure 4

Stress relaxation curves for various values of fixed stretch for plasticized polyvinyl chloride (40% tricresyl phosphate) [8]. Notice that the curve for a stretch of 1.30 crosses the curves for small stretch.

Figures 5 and 6 show the results of double-step stress relaxation experiments on a high molecular weight polyisobutylene (Vistanex L-100) [10]. The first step was held for 80 minutes in the experiment of figure 5. and for 60 minutes in that of figure 6. In each figure the data are plotted in two curves taking as zero the instants of the first and second steps, respectively. This way of plotting emphasizes the critical data at times just after the steps. The solid black points represent values calculated with data of single-step stress relaxation. Figure 7 shows the results for this same polyisobutylene of a triple-step stress relaxation experiment in which the first step was held for 20 minutes, the second for 40 minutes. Some small deviations are seen at early times immediately after the steps but, on the whole, the model appears to be quite accurate.

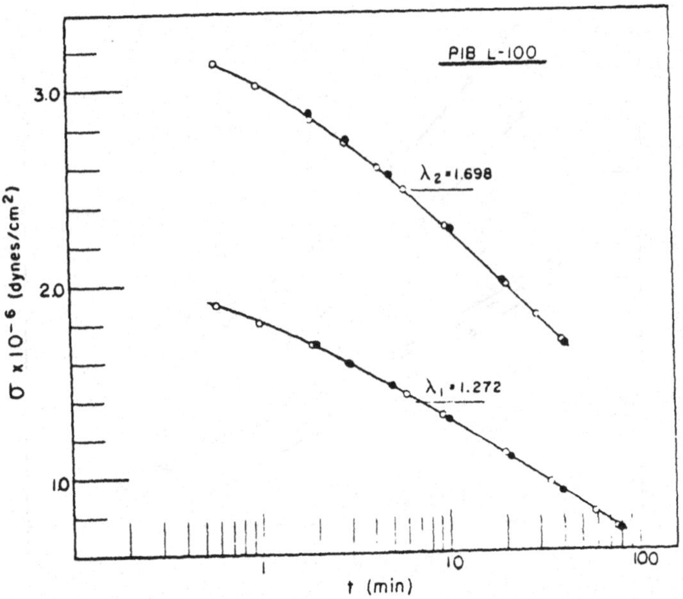

Figure 5
Open circles are data of a double-step stress relaxation of polyisobutylene [10]. The filled circles are points calculated from single-step stress relaxation by use of the superposition of equation (21).

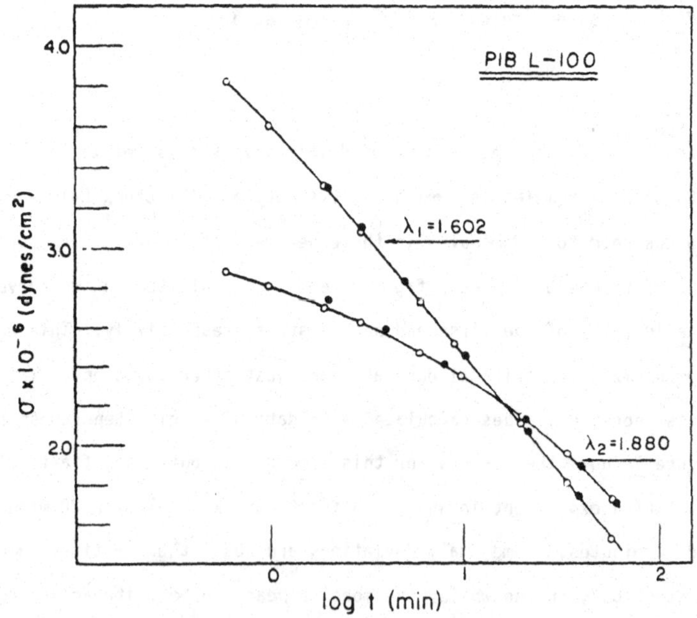

Figure 6
A double-step stress relaxation similar to that of figure 5 but for somewhat greater stretches [10]. The data of the second stretch shows slight deviations from the results predicted from single-step stress relaxation data.

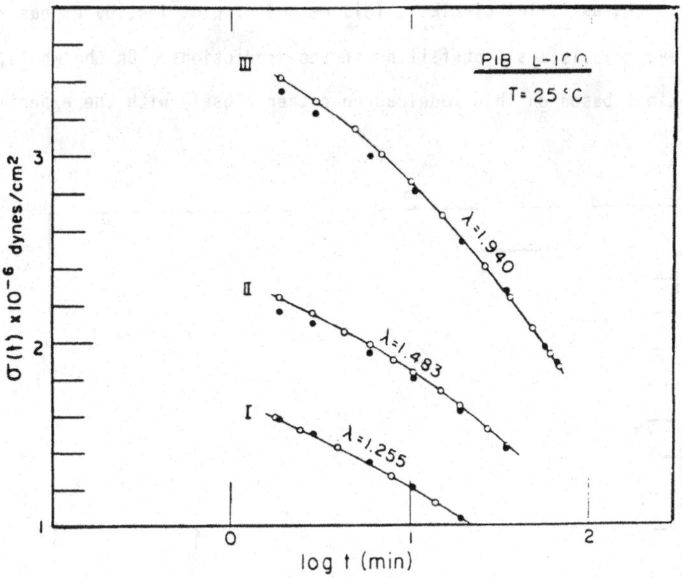

Figure 7

A triple-step stress relaxation for the same material as shown in figure 5 and 6. [10]. Deviations from the predicted results are small and were considered to be within experimental accuracy.

Figure 8 shows the results of an experiment in creep and recovery of polyisobutylene [10]. The experimental values of stretch are plotted as open circles and the values calculated from the stretch history through the use of single-step stress relaxation data are plotted as filled circles. Figure 9 is a replot of the curve of stretch history during which creep occurs. Values of stretching force have been calculated through equation (29) and indicated at various points along the curve. The actual stretching force was 86 grams and it is seen that the discrepancies were not much more than one percent. Figure 10 is a replot of the recovery part of the stretch history. The instantaneous stretch is plotted against the logarithm of the time interval from the instant of removing the stretching force. Again, measured values are plotted as open circles and calculated values as filled circles. This method of plotting, by emphasizing the discrepancies, reveals a slight failing of the predictions. On the whole, however, the calculations based on this model agree rather closely with the experimental measurements.

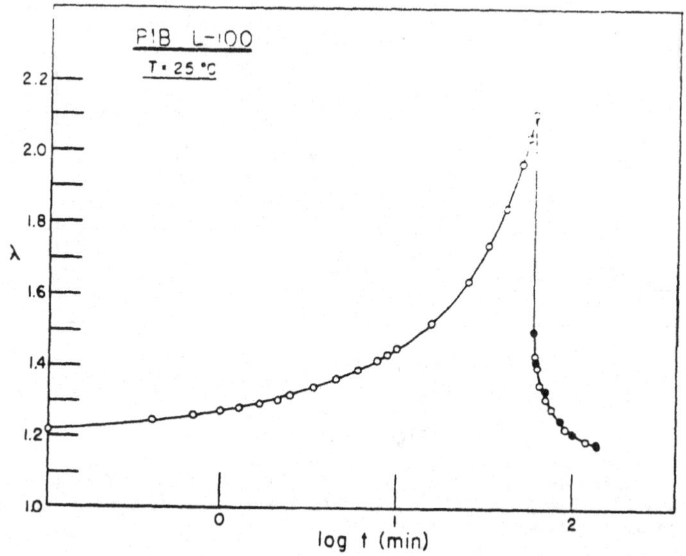

Figure 8

Creep and recovery of polyisobutylene [10]. Open circles mark points for measured data and filled circles are points calculated from stress relaxation data.

The double-step superposition principle of equation (21) is valid for deformations in shearing as well stretch if one takes into account the fact that shear strains compound by adding rather than by multiplying as stretches. McKenna and Zapas noticed an interesting simplification of the superposition which applied for two-step stress relaxation in shear when the second step is opposite in sign and one half the magnitude of the first step [20]. Because the shear stress must be an odd function of shear strain while the normal stress differences must be even functions of shear strain, the superposition of equation (21) can be written in the following surprisingly simple form:

$$\sigma(t) = 2\,\sigma_{SR}(\gamma, t + t_1) - \sigma_{SR}(\gamma, t),$$
$$\nu(t) = \nu_{SR}(\nu, t), \tag{30}$$

where $\sigma(t)$ is the shear stress and $\nu(\tau)$ is the first normal stress difference for the instant t after the second step. The first step is a shear of magnitude 2γ which occurs at instant $-t_1$ and the second step is of magnitude $-\gamma$ and occurs at instant zero. In this special case the shear stress relaxation can be expressed with data of only two single-step relaxations but, even more surprising, the normal stress is in fact equal to a single-step relaxation! That is to say, the normal stress is independent of the duration of the first step.

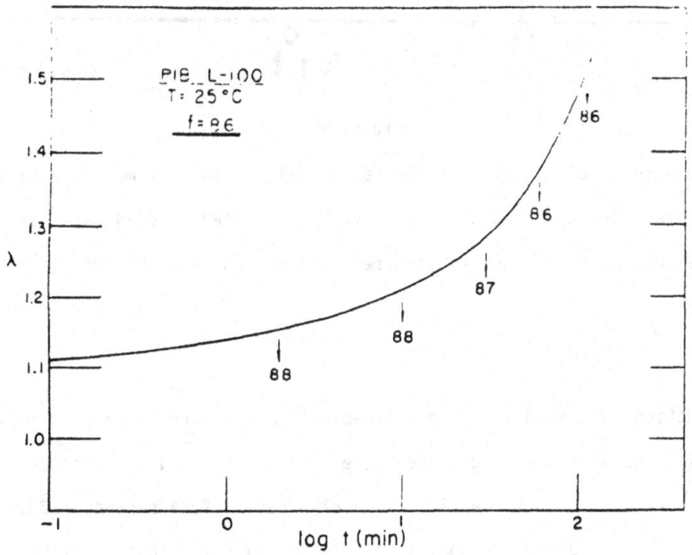

Figure 9

The creep curve of figure 8 replotted. The stretching force as calculated from stress relaxation data is shown by arrows at various points along the curve. The actual value was 86 grams.

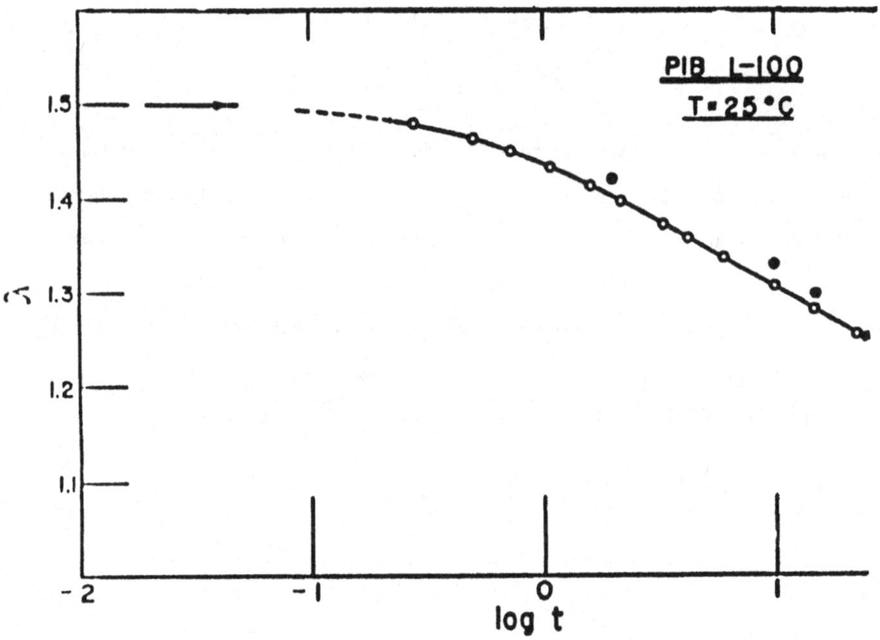

Figure 10

The recovery curve of figure 8 replotted on a logarithmic time scale based on the instant of release of the stretching force. The filled circles mark values of stretch calculated from single-step stress relaxation data through a numerical iteration.

This prediction is tested for a polymer solution in figure 11, a plot in which the squares mark the measured single-step stress relaxation and the circles mark the double-step relaxation after a first step of duration 2.44 seconds [22]. The two sets of data do indeed appear to lie on a single curve. Figure 12 is another test of this prediction for a much different material, polymethylmethacrylate or "plexiglass". The points of this plot show the normal stress for double-step relaxation at two widely spaced isochrones for durations of the initial step ranging over three decades of time. The solid lines indicate the constant values predicted from measurements of single-step stress relaxation and the error bars indicate that the data agree to within the standard deviations corresponding to the dotted lines. The agreement of the normal stress data with this surprising prediction is reassuring.

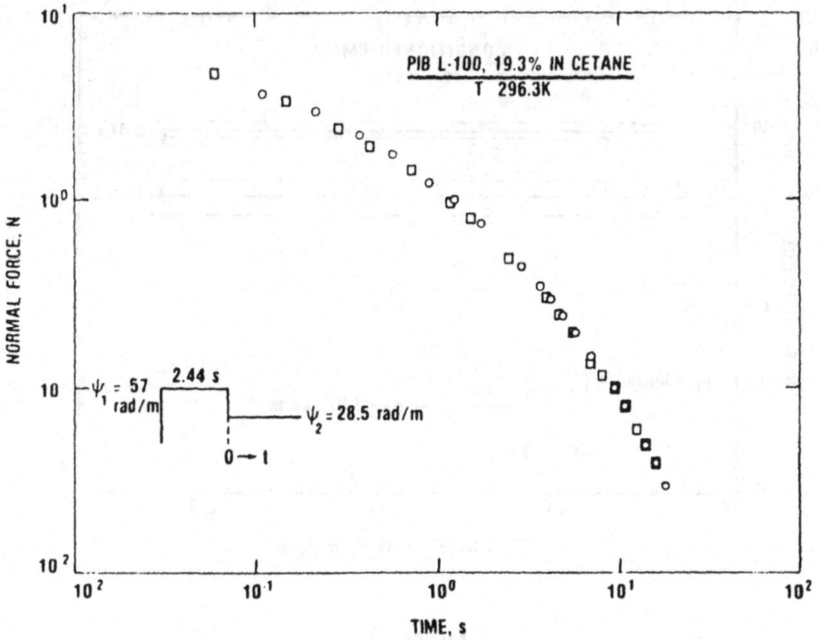

Figure 11

Normal force response of a double-step stress relaxation in torsion (circles compared with a single-step relaxation (squares) for a polyisobutylene solution [22]. The magnitude of the second step is half the magnitude of the initial step and the two sets of data superpose.

The predictions of equation (30) for shear stress relaxation are tested in figure 13. In this figure the results of using single-step relaxation data with the superposition principle are plotted as open circles while actual measurements of double-step relaxation are plotted as filled circles. The plot has been arranged to emphasize the discrepancies which would appear small if compared to the stresses required to produce the initial step. Nevertheless, the discrepancies are almost certainly real since they exceed a reasonable estimate of the experimental errors. What is more, similar results have been obtained by independent experiments using somewhat different techniques [24].

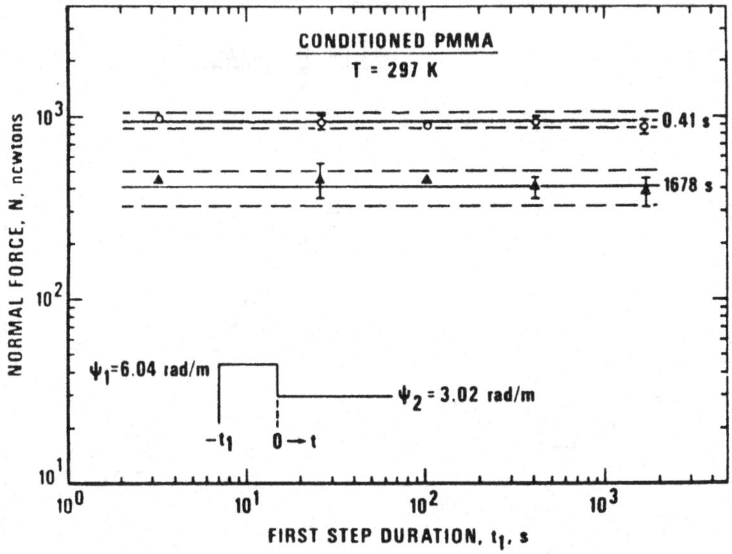

Figure 12

Normal stress data from double-step stress relaxation experiments where the second step is half the initial step [22]. The data indicate that the relaxation after the second step is independent of the duration of the first step.

For very fluid materials, equation (29) can be more easily tested with another type of deformation history which might be called "suddenly applied steady shearing". In this test, the fluid is loaded into a rheogoniometer and allowed to relax to a stress-free equilibrium. Then, as instantaneously as possible, a shearing of constant rate is induced and the shear stress and normal stress are monitored.

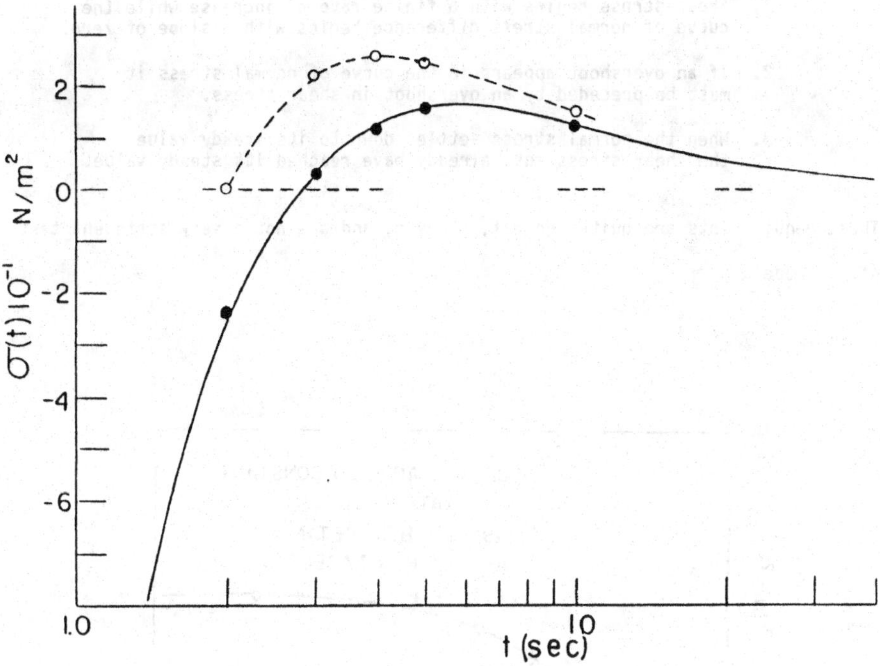

Figure 13

Shear stress data for the experiment of figure 11. The open circles are calculated from single-step stress relaxation data. The choice of plot emphasizes the discrepancies.

Figure 14 shows some data which was taken in this way using a polymer solution [18]. The shear stress clearly has an "overshoot" in this figure, that is, it rises to a maximum before relaxing to a steady value. The normal stress also would have shown an overshoot if the plot had extended beyond 4 seconds. The following qualitative features associated with the model of equation (29) are easily verified in this figure:

1. Shear stress begins with a finite rate of increase while the curve of normal stress difference begins with a slope of zero.

2. If an overshoot appears in the curve of normal stress it must be preceded by an overshoot in shear stress.

3. When the normal stress settles down to its steady value the shear stress must already have reached its steady value.

These requirements are quite general, however, and are not a very stringent test of the model.

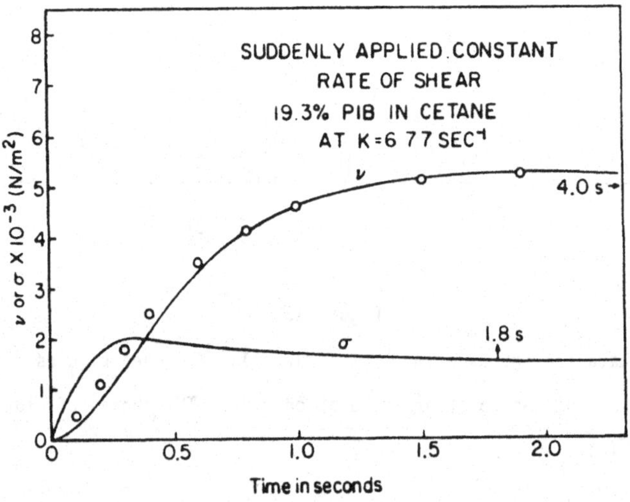

Figure 14

Sample data of suddenly applied constant shearing [18]. The normal stress curve did not reach an equilibrium value until 4 seconds, a time beyond the range of the plot. "Overshoot" occurs in both curves. The open circles are values of normal stress calculated from the curves of shear stress. It is probable that the discrepancies reflect errors in measured normal stress at early times.

A more quantitative test of equation (29) for fluids undergoing suddenly applied steady shearing can be made through some so-called "rheological relations" between the history of the shear stress and the history of the first normal stress difference [18]. The relations are most easily expressed through an auxiliary quantity, $\Gamma(\kappa,t)$ defined as follows:

$$\Gamma(\kappa,t) \equiv t\sigma(\kappa,t) - \int_0^t \sigma(\kappa,\tau)d\tau \tag{31}$$

where κ is the constant rate of shear. This quantity $\Gamma(\kappa,t)$ can be visualized as an area in a plot, for shear rate κ, of shear versus time. Figure 15 is such a plot and $\Gamma(\kappa,t)$ is represented by the difference between area A and area B. In the limit of long times, Γ will approach a steady state value independent of time. We indicate this steady state value with a subscripted S and write:

$$\Gamma_S(\kappa) = \int_0^\infty [\sigma_S(\kappa) - \sigma(\kappa,\tau)]d\tau. \tag{32}$$

The steady state normal stress is related to this quantity by the following relation:

$$\kappa\, \Gamma_S(\kappa) = \nu_S(\kappa) - \int_0^\kappa \frac{\nu_S(\xi)}{\xi}\, d\xi$$

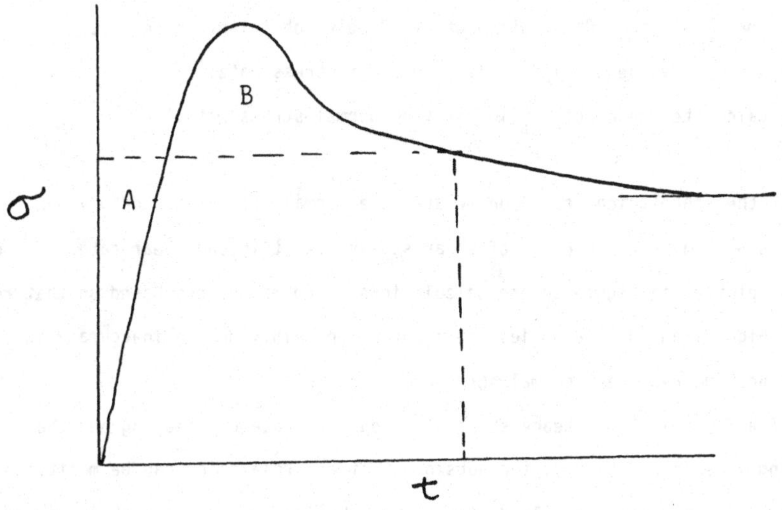

Figure 15

Schematic graphically illustrating the calculation of the quantity $\Gamma(\kappa,t)$, the area of the rectangle marked out with dotted lines minus the area under the curve. By eliminating the overlap of these two areas it can be seen that Γ is just area A minus area B.

Table I displays values of $\Gamma_S(\kappa)$ measured through the use of equation (32) and calculated through the rheological relation, equation (33), for various values of shear rate. In all cases the discrepancies could be explained by an error in the steady value of the shear stress of 3% or less.

Table I[a,b]

$$\kappa\Gamma_S(\kappa) = \nu_S(\kappa) - \int_0^{\kappa} \frac{\nu(\xi)}{\xi} d\xi$$

κ (shear rate)	0.352	0.702	1.40	2.80	5.56	11.1	22.2
Measured[c] $\Gamma_S(\kappa)$	16.0	3.04	11.7	6.38	2.00	- 3.46	-10.0
Calculated[d] $\Gamma_S(\kappa)$	10.7	4.98	5.64	11.0	4.98	- 0.405	- 4.32
$\Delta\Gamma_S$ due to 1% error in $\sigma_S(\kappa)$	2.0	1.5	2.5	3.8	1.6	4.5	1.9

[a] All quantities are in Standard International units.

[b] Test material is a 10% solution of L-100 polyisobutylene in cetane.

[c] The measured value of $\Gamma_S(\kappa)$ is from shear-stress data.

[d] The calculated value of $\Gamma_S(\kappa)$ is from normal-stress data.

Another rheological relation relates the normal stress in suddenly applied steady shearing to an integral of shear stress over time and shear rate. The open circles plotted in figure 14 are calculations of normal stress based on that relation. Significant discrepancies are apparent, possibly due to inaccuracy in measuring time dependent normal stresses.

If a fluid is in a steady state at a constant rate of shearing and the shearing is suddenly halted, the subsequent stress relaxation can be monitored. In this case, equation (29) leads to another rheological relation which may be checked by experiment [18], viz.:

$$\int_0^{\kappa} \frac{\nu_S(\xi)}{\xi} d\xi = \int_0^{\infty} \sigma(\kappa, \tau) d\tau, \tag{34}$$

where σ is the monitored stress.

Table II[a,b]

$$\int_0^\kappa \frac{\nu_s(\xi)}{\xi}\,d\xi = \kappa \int_0^\infty \sigma(\kappa,\tau)\,d\tau$$

κ (shear rate)	0.177	0.556	1.11	1.77	5.56	11.1
$\kappa \int_0^\infty \sigma\varepsilon(\kappa,\tau)d\tau$ $(\times 10^3)$	0.133	0.634	1.20	2.05	6.17	15.1
$\int_0^\kappa \frac{\nu_s(\xi)}{\xi} d\xi$ $(\times 10^3)$	0.133	0.610	1.30	2.05	6.12	11.3
$\frac{1}{2}\nu_s(\kappa)$ $(\times 10^3)$	0.115	0.365	0.59	1.05	2.34	3.22

[a] All quantities are in Standard International Units

[b] Test material is a 19.3% solution of L-100 polyisobutylene in cetane.

Table II displays the two sides of this equation as calculated from data on a solution of polyisobutylene. The agreement generally is well within the accuracy of the data with perhaps a slight question at the highest shear rate.

The third row of table II relates to constitutive equations of a slightly different form. If in place of the derivatives of U scalar functions of the rate of deformation tensor occurred, the resulting model would require that the third row of table II be equal to the first row. The data clearly do not support such a model, but they do support the model of equation (29).

There is a considerable body of data not cited here to be found in the literature which bears on the questions of the applicability of the BKZ model (and other models) to a variety of materials. Examples can be found either to confirm or to deny the validity of the model. Perhaps it is appropriate to bear in mind a quotation taken from the writings of Eddington:

> "It is also a good rule not to put too much confidence in experimental results until they have been confirmed by theory."

Experimental results are always based upon a great number of tacit assumptions. For instance, in most of the results we quote above, we have deliberately ignored the inertia of the materials since the experiments concerned were in conditions where that is considered justified. Other much more subtle experimental errors can occur. Figure 16 shows shear stress buildup for suddenly applied steady shearing which was measured using torsion bars of differing stiffness in the same rheogoniometer [16]. The differences are rather surprising and illustrate the stringent necessity for surprisingly stiff torsion bars.

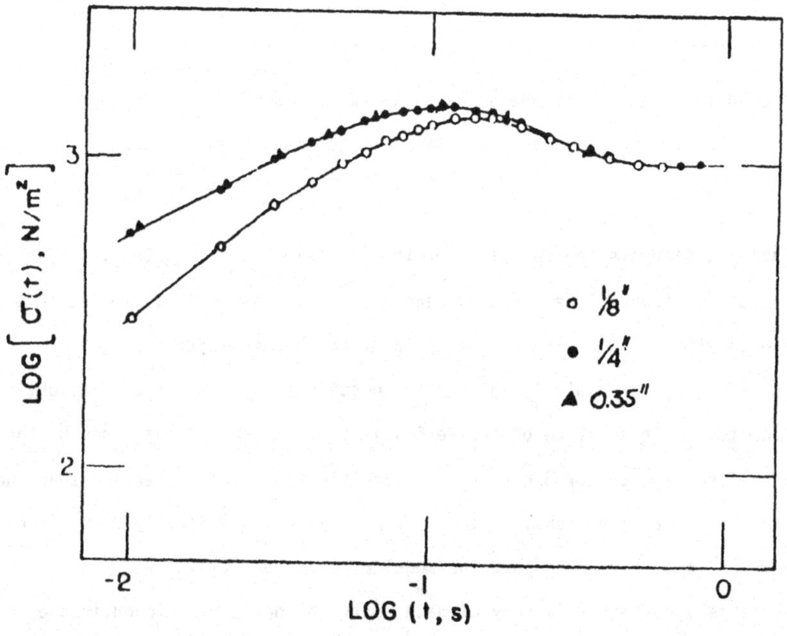

Figure 16

The stiffness of the torsion bar of a rheometer can affect the measuremenet of shear stress as a function of time in suddenly applied constant shearing [16].

Figure 17 shows normal stress measurements taken with cone and plate geometry in a rheogoniometer [18]. A dependence on cone angle is evident. It is comforting that in these particular cases the measurements seem to be reasonably consistent for sufficiently stiff torsion bars or large enough cone angles (but not too large). One recalls, though, the notorious history of measurement of normal stresses with "stand pipes". Although the measurement method was not completely understood, experimenters felt secure in the knowledge that for sufficiently small holes the measurements were independent of hole size. Only after many years was it demonstrated that there was an intrinsic error in the method [15].

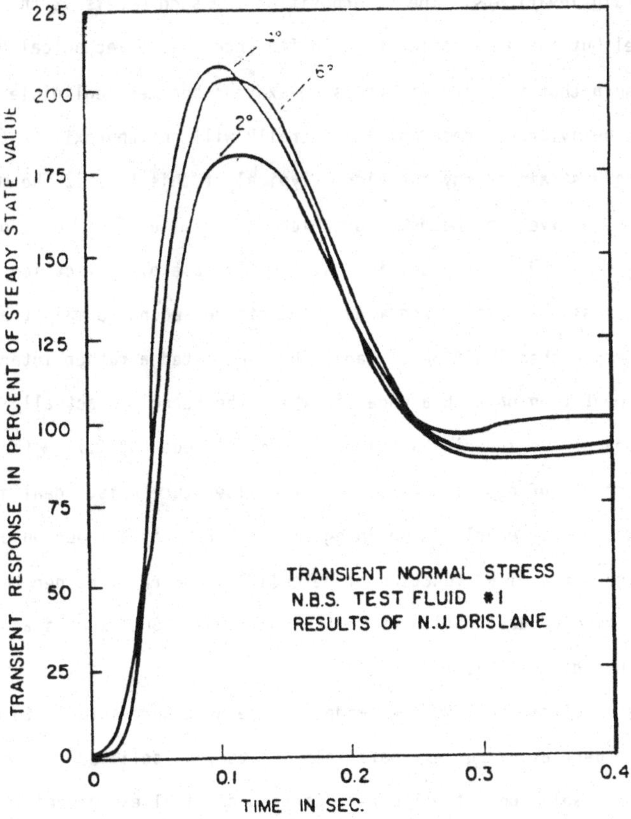

Figure 17

The cone angle in a cone-and-plate rheometer can affect the measurement of transient normal stress [18]. Insufficient axial or torsional stiffness of the machine can also introduce serious errors.

There are, of course, innumerable constitutive equations in the rheological literature, each with its own useful features and its own failings. A model based on a detailed molecular picture which is currently of great interest is the Doi-Edwards model [19]. This model, at least in its simplest form, results in a special type of BKZ model with a kernel function which is a product of a time dependent factor and a strain dependent factor. It is therefore subject to all the failings of the BKZ model. A later version of the Doi-Edwards model leads to a non-BKZ form and a modified superposition principle [24]. There are a number of models which are very close to BKZ models in that they are of the form of equation (29) except that the derivatives of U are replaced by various functions of the relative strain invariants. The superposition for such models is the same as for the BKZ model but there is an important difference in the mechanical behavior. In situations such that the rate of stress relaxation becomes negligible at finite strains, the behavior of these models generally will not approximate that of a material with a strain energy function. Typical of this type is the model of Papanastasiou, Scriven and Macosko for which the term in C inverse of the kernel is taken as zero [25]. The model is designed for ease of application but as a price for this simplification the model predicts no second normal stress difference in simple shearing flows. Wagner has suggested a rather interesting further ornamentation of such a model in which the kernel is actually a functional of the history of the invariants [21]. This model does not follow the superposition principle of equation (21). A large literature exists dealing with models for which the kernel of the integral for stress is a function of the rate of deformation, or a mixed function of deformation and rate, rather than simply a function of deformation as in the BKZ model. Such models are not considered in this discussion.

For most applications, the BKZ model can be used successfully to describe the mechanical properties of rubbery materials, be they as solid as vulcanized rubber or as liquid as solutions of polymers. Certain critical experiments, however, can emphasize failings of the model. The successes and the failures of the BKZ model can be used as guides and tests in examining the validity of models based on considerations of molecular structure.

Bibliography

1. Flory, Network structure and the elastic properties of vulcanized rubber, Chem Rev. 35, 51, (1944).

2. M.S. Green and A.V. Tobolsky, A new approach to the theory of relaxing polymeric media, Journ. Chem. Phys. 14, 80, (1946).

3. L.R.G. Treloar, The Physics of Rubber Elasticity, Oxford University Press, London, 1949.

4. M. Yamamoto, The viscoelastic properties of network structure I. General formulation, Journ. Phys. Soc. Japan 11, 413, (1956).

5. A.S. Lodge, A network theory of flow birefringence and stress in concentrated polymer solutions, Trans. Faraday Soc. 52, 120, (1956).

6. E.A. DiMarzio, Contribution to a liquid-like theory of rubber elasticity, Journ. Chem. Phys. 36, 1563, (1962). See also ACS Preprints 9, 256, (1968).

7. A. Kaye, Non-Newtonian flow in incompressible fluids, Part I: A general rheological equation of state; Part II: Some problems of steady flow; Part III: Some problems in transient flow, Technical Note 134, College of Aeronautics, Cranfield, U.K., (1962).

8. B. Bernstein, E.A. Kearsley and L.J. Zapas, A study of stress relaxation with finite strain, Trans. Soc. Rheology VII, 391, (1963).

9. B. Bernstein, E.A. Kearsley and L.J. Zapas, Thermodynamics of perfect elastic fluids, Journ. Research National Bureau of Standards 68B, 103, (1964).

10. L.J. Zapas and T. Craft, Correlation of large longitudinal deformations with different strain histories, Journ. Research National Bureau of Standards 69A, 541, (1965).

11. C. Truesdell and W. Noll, The Non-linear Field Theories of Mechanics, Springer-Verlag, Berlin and New York, 1965.

12. B. Bernstein, Time-dependent behavior of an incompressible elastic fluid-Some homogeneous deformations, Acta Mechanica II, 329, (1966).

13. L.J. Zapas, Viscoelastic behavior under large deformations, Journ. Research National Bureau of Standards 70A, 525, (1966).

14. K.C. Valanis and R.F. Landel, The strain-energy function of a hyperelastic material in terms of the extension ratios, Journ. Appl. Phys. 38, 2997, (1967).

15. J.M. Broadbent, A. Kaye, A.S. Lodge and D.G. Vale, Possible systematic error in the measurement of normal stress differences in polymer solutions in steady shear flow, Nature 217, 55, (1968).

16. L.J. Zapas and J.C. Phillips, Simple shearing flows in polyisobutylene solutions, Journ. Research National Bureau of Standards 75A, 33, (1971).

17. L.J. Zapas, Non-linear behavior of polyisobutylene solutions, in Deformation and Fracture of High Polymers, H.H. Kausch editor, Plenum, 1974.

18. E.A. Kearsley and L.J. Zapas, Experimental tests of some integral rheological relations, Trans. Soc. Rheology 20, 623, (1976).

19. M. Doi and S.F. Edwards, Dynamics of concentrated polymer systems, Part 1. Brownian motion in the equilibrium state; Part 2. Molecular motion under flow; Part 3. The constitutive equation, Journ. Chem. Soc. Faraday Trans. II 74, 1789; 1802; 1818, (1978), Part 4. Rheological properties, Journ. Chem. Soc. Faraday Trans. II 75, 38, (1979).

20. G.B. McKenna and L.J. Zapas, Nonlinear viscoelastic behavior of poly(methyl methacrylate) in torsion, Journ. Rheology 23, 151, (1979).

21. M.H. Wagner, Zur netzwerktheorie von Polymer-Schmelzen, Rheol. Acta 18, 33, (1979) also (with J. Meissner), Network disentanglement and time-dependent flow behavior of polymer melts, Makromol. Chem 181, 1533, (1980).

22. G.B. McKenna and L.J. Zapas, The normal stress response in nonlinear viscoelastic materials: Some experimental findings, Journ. Rheology 24, 367, (1980).

23. L.J. Zapas and J.C. Phillips, Nonlinear behavior of polyisobutylene solutions as a function of concentration, Journ. Rheology 25, 405, (1981).

24. K. Osaki, S. Kimura and M. Kurata, Relaxation of shear and normal stresses in double-step shear deformations for a polyisobutylene solution. A test of Doi-Edwards theory for polymer rheology, Journ. Rheology 25, 549, (1981).

25. A.C. Papanastasiou, L.E. Scriven and C.W. Macosko, An integral constitutive equation for mixed flows: Viscoelastic characterization, Journ. Rheology 27, 387, (1983).

26. E.A. Kearsley, Asymmetric stretching of a symmetrically loaded elastic sheet, Internat. Journ. Solids and Structures, to appear.

DYNAMICAL BEHAVIOR UNDER RANDOM PERTURBATION OF MATERIALS WITH SELECTIVE RECALL

Moshe Marcus

Technion
Haifa, Israel

Victor J. Mizel*

Department of Mathematics
Carnegie-Mellon University
Pittsburgh, PA 15213

Introduction

In this work we propose a new formulation of stochastic functional differential equations which, in contrast to the standard formulation of such equations, leads to a non-degenerate diffusion process.

Consider a deterministic f.d.e. problem of the form,

$$\frac{dx}{dt} = g(t,x(t),x^t) \quad \text{for} \quad t > 0, \quad x^0 = \phi, \quad x(0) = \alpha, \tag{0.1}$$

where α is a real number, ϕ is an element in a "history space" H consisting of functions on $(0,\infty)$ (for instance a weighted L_2 space) and x^t is the function on $(0,\infty)$ defined by $x^t(a) = x(t-a)$. Accordingly, g is a functional on $R^+ \times R \times H$ or on an appropriate subspace thereof.

The standard way of formulating a stochastic version of (0.1) is to replace the given equation by

$$dx = g(t,x(t),x^t)dt + k(t,x(t),x^t)dw \tag{0.2}$$

where w is a Wiener process. Stochastic functional differential equations of this form have been treated in [K], [Mo1, 2] in the context of continuous histories of finite duration, and in [MT] in the context of weighted L_p spaces on $(0,\infty)$. In these papers, existence and uniqueness of solutions was established for fairly general classes of equations of the form (0.2). In addition [K] and [MT] obtain stability results (via Lyapunov functionals) in some special cases.

---

* Research partially supported by the National Science Foundation and the Air Force Office of Scientific Research

In the case of an ordinary stochastic d.e., it is well known that important aspects of the stochastic process associated with the equation are closely related to the properties of the infinitesimal generator of the corresponding Markov process. For instance the stability in probability of a solution is closely related to the existence of certain Lyapunov functions v which (among other conditions) must satisfy the inequality $Lv \leq 0$ where L is the above-mentioned infinitesimal generator (a parabolic partial differential operator). This relation holds (in part) even if the parabolic operator is singular, as is the case for equations of the form (0.2). However, stronger concepts of stability (such as asymptotic stability in probability) depend very strongly on the non-degeneracy of the infinitesimal generator. Furthermore, the construction of Lyapunov functions is greatly facilitated when the infinitesimal generator is non-degenerate, (see [H; Ch.5]). Another concept of stability is related to the concept of expected exit time, and the latter can be obtained as a solution of an initial-boundary value problem for the equation $Lv = -1$, with L as before (see, [S2,3]). Clearly the existence of such solutions is much more easily established when L is non-degenerate.

If one derives the infinitesimal generator of (0.2) (for $H = L^2_\rho(0,\infty)$, a weighted L_2 space) one obtains a highly singular parabolic operator, namely,

$$Lu = \frac{\partial u}{\partial t} + g\frac{\partial u}{\partial \alpha} + \frac{1}{2}k^2\frac{\partial^2 u}{\partial \alpha^2} + \lim_{h\to 0+} \left\langle \frac{\partial u}{\partial x}, \frac{\hat{T}^h(\alpha,x) - x}{h} \right\rangle_{L^2_\rho} \tag{0.3}$$

where $u = u(t,\alpha,x)$ is a functional on $R^+ \times R \times H$, $\frac{\partial u}{\partial x}$ denotes a Frechet derivative and

$$\hat{T}^h(\alpha,x) = \begin{cases} \alpha & \text{for } 0 < a \leq h \\ x(a-h) & \text{for } h < a. \end{cases}$$

A similar formula is obtained in [Mo1] in the case where H is the space of continuous functions on a bounded interval $[0,\ell]$ (cf. also [MT]).

Note that the coefficient of $\frac{\partial^2 u}{\partial x^2}$ in (0.3) is zero. This stems from the fact that (0.2) contains only a one-dimensional diffusion term while the dependent variable $(x(t),x^t)$ is in the infinite dimensional space $\mathbb{R} \times H$.

Our formulation of a stochastic version of (0.1) is based on a different interpretation of the deterministic f.d.e., which we describe below (see also [MM1,2]).

Consider the following problem, in which X is a function of two variables $(t,a) \in [0,T] \times [0,\infty]$ and $\overline{X}(t) := (X(t,0),X(t,\cdot))$,

$$(0.4) \quad \begin{cases} \frac{\partial X}{\partial t} + \frac{\partial X}{\partial a} = 0 \text{ for } 0 < t, \quad 0 < a < \infty; \quad X(0,\cdot) = \phi(\cdot) \\ \frac{\partial X}{\partial t}(t,0) = g(t,\overline{X}(t)), \qquad 0 < t; \qquad X(0,0) = \alpha . \end{cases}$$

Since the general solution of equation $(0.4)_1$ is of the form $X = x(t - a)$ it is easily seen that (0.4) is equivalent to (0.1) with $x(t) = X(t,0)$.

The first equation in (0.4) specifies how the history is to be interpreted before it is inserted into the second equation, which determines the immediate evolution of the process. In the case described in (0.4) the perceived history at time $t + \Delta t$ is an <u>extension</u> of the history at time t shifted Δt units in a. This extension is determined by the evolution of the process from time t to $t + \Delta t$. If, however, we replace $(0.4)_1$ by the equation

$$(0.5) \qquad \frac{\partial X}{\partial t} + \frac{\partial X}{\partial a} = F(t,a,\overline{X}(t))$$

then the perceived history at time $t + \Delta t$ is a <u>modification</u> of the history at time t (shifted and extended as before). That is, in (0.5) the perception of history changes as history shifts in time.

Starting from (0.4) we shall formulate the stochastic version of the f.d.e. (0.1) as follows:

$$(0.6) \quad \begin{aligned} &DX = \beta(t,a,\overline{X}(t))DW, \quad 0 < t, \ 0 < a; \ X(0,\cdot) = \phi(\cdot) \\ &dX(t,0) = g(t,\overline{X}(t))dt + k(t,\overline{X}(t))dw; \ X(0,0) = \alpha. \end{aligned}$$

Here D denotes the differential along the characteristic line $t - a = \text{const.}$, W is a two dimensional Wiener process in the quarter plane $0 < t$, $0 < a$ (see e.g.

[CW]) and w is a one-dimensional Wiener process independent of W. Thus the deterministic term in (0.5) is here replaced by a white noise term, namely βDW. This term expresses an element of uncertainty in the perception of history. In addition there is, of course, the white noise term kdw which expresses a random perturbation in the mechanism acting in the present.

In general one may allow the first equation in (0.6) to contain a drift term as well as a diffusion term. Of course in that case (0.6) can no longer be considered as a purely random perturbation of (0.1). However it can be related to the more general equations discussed in [MM1,2].

In this note we shall discuss problem (0.6) in the case where β is independent of $\overline{X}$. The questions that will be discussed are: existence and uniqueness of solutions, the Markov property of the solution process and the representation and other properties of the (weak) infinitesimal generator of this process. The main results will be presented without proof. Detailed proofs and related results will appear elsewhere.

1. The Existence Theorem

Consider the system of equations:

$$
\text{(E)} \qquad \begin{aligned}
DX(t,a) &:= X_{,t}(t,a) + X_{,a}(t,a) = \beta(t,a)DW(t,a) \\
dX(t,0) &= g(t,\overline{X}(t))dt + k(t,\overline{X}(t))dw \\
X(t_0,0) &= \phi_0, \quad X(t_0,\cdot) = \phi \in L_2^\rho(0,\infty),
\end{aligned}
$$

where $W : \mathbb{R}_+ \times \mathbb{R}_+ \times \Omega \to \mathbb{R}$ and $w : \mathbb{R}_+ \times \Omega \to R$ are, respectively, a two parameter standard Brownian motion and an independent one-parameter standard Brownian motion, and $L_2^\rho(0,\infty)$ is a fading memory space in the sense of [CM1,2]. (The precise conditions that we require on the weight ρ are listed further below).

As in [CW], W is defined by

$$
W(t^*,a^*) = \overline{W}(0 < t \leq t^*,\ 0 < a \leq a^*) \qquad \forall (t^*,a^*) \in R_+^2,
$$

where $\overline{W}$ is a random measure in $\mathbb{R}_+^2$ which assigns to each Borel set A a

Gaussian random variable of mean zero and variance $m(A)$ (the Lebesgue measure of A) and which assigns independent random variables to disjoint sets.

In the following we set

$$\hat{g}(t) = g(t,\overline{X}(t)), \quad \hat{k}(t) = k(t,\overline{X}(t)).$$

Recall that $\overline{X}(t) = (X(t,0),X(t,\cdot))$. The system (E) will be interpreted in integrated form:

$$(E_s) \qquad X(t,a) = \begin{cases} X(t-a,0) + \int_{t-a}^{t} \beta(r,a-t+r)\partial_r W(r,a-t+r) \\ \qquad\qquad \text{in Dom I} := \{(t,a) : a < t - t_0\}, \\ \phi(a+t_0-t) + \int_{t_0}^{t} \beta(r,a-t+r)\partial_r W(r,a-t+r) \\ \qquad\qquad \text{in Dom II} := \{(t,a) : a > t - t_0\}. \end{cases}$$

$$X(t,0) = \phi_0 + \int_{t_0}^{t} \hat{g}(r)dr + \hat{k}(r)dw(r), \quad t > t_0.$$

We shall use the following σ-algebras below:

$$\mathcal{G}_{t_0} := \sigma\{\phi_0,\phi(a);\ a > 0\} \text{ and } \mathcal{G}_t := \mathcal{G}_{t_0} \vee \mathcal{F}_t$$

where

$$\mathcal{F}_t := \sigma\{w(u) - w(t_0),\ W(u,b) - W(t_0,a);\ u \in [t_0,t],\ b > a > 0\}.$$

<u>Definition</u>. We say that the two-parameter stochastic process $\{X(t,a) : t \geq t_0, a > 0\}$ and the one-parameter stochastic process $\{X(t,0) : t > t_0\}$ are a (stochastically strong) solution of (E) in $[t_0,T]$ if:

1. The processes are adapted to $\mathcal{G}_t$ for $\forall t \in [t_0,T]$,
2. With probability one,

$$\overline{X}(t) := (X(t,0),X(t,\cdot)) \in \mathbb{R} \times L_2^{\rho}(0,\infty) =: \underline{X} \quad \forall t \in [t_0,T]$$

3. $\hat{g} \in L_1(t_0,T)$, $\hat{k} \in L_2(t_0,T)$ a.s.
4. $\overline{X}(t,0) = \overline{\phi} := (\phi_0,\phi)$ a.s.
5. With probability one, (E_s) holds for each t in $[t_0,T]$.

Assumptions on ρ

In this paper it will always be assumed that ρ satisfies the following conditions:

a) $\rho \in L_1(0,\infty)$ and $\rho > 0$ everywhere

b) $\overline{K}(\sigma) := \operatorname{ess\,sup}_{a \in \mathbb{R}_+} \frac{\rho(a+\sigma)}{\rho(a)} < \infty, \quad \forall \sigma > 0$

c) $\overline{K}(\sigma)$ is locally bounded in $[0,\infty)$.

Notation

1. If $\overline{\psi} = (\psi_0, \psi) \in \underline{X} := \mathbb{R} \times L_2^{\rho}(0,\infty)$, we denote,

$$\|\overline{\psi}\|_{\underline{X}}^2 = |\psi_0|^2 + \|\psi\|_{L_2^{\rho}}^2 \quad .$$

2. Given a weight function ρ we denote,

$$\beta(\alpha) = \begin{cases} \rho(\alpha) & \text{if } \alpha > 0 \\ 1 & \text{if } \alpha < 0 \end{cases} \quad .$$

Theorem A. Suppose $g,k : \mathbb{R}_+ \times \underline{X} \to R$ satisfy, for some $L > 0$,

$$\text{(L)} \qquad \begin{cases} |g(t,\overline{\phi}) - g(t,\overline{\psi})| + |k(t,\overline{\phi}) - k(t,\overline{\psi})| \leq L\|\overline{\phi} - \overline{\psi}\|_{\underline{X}} \\ |g(t,\overline{\psi})| + |k(t,\overline{\psi})| \leq L(1 + \|\overline{\psi}\|_{\underline{X}}), \end{cases}$$

for all $\overline{\phi},\overline{\psi} \in \underline{X}$ and all $t \in [t_0,T]$.

Further, suppose that $\beta : \mathbb{R}_+ \times \mathbb{R} \to \mathbb{R}$ is Lebesgue measurable and satisfies:

(H)

$\beta(r,b) = 0$ for $b < 0$ and,

for every $\alpha \in R$, the function $\beta(\cdot,\cdot + \alpha) \in L_2(0,T)$.

The function $\mathbb{R} \ni \alpha \mapsto \beta(\cdot,\cdot+\alpha) \in L_2(0,T)$ is continuous.

$$\int_0^\infty \int_0^T (1+\alpha)\beta^2(r,r+\alpha)\,dr\,\rho(\alpha)\,d\alpha < \infty.$$

Finally, suppose that $\overline{\phi} : \Omega \to \underline{X}$ is independent of $\mathcal{F}_t (\forall t \geq t_0)$ and that

$E\|\bar{\phi}\|^2_{\underline{X}} < \infty$. Then there exists a unique stochastically strong solution of (E) in $[t_0,T]$. The solution $\bar{X}$ satisfies the boundedness condition,

$$\text{(i)} \qquad \sup_{t_0 \le t \le T} E(\|\bar{X}(t)\|^2_{\underline{X}}) = M_{\bar{\phi}} < \infty,$$

and for every positive c, $M_{\bar{\phi}}$ is uniformly bounded for those $\bar{\phi}$ in $\underline{X}$ such that $E(\|\bar{\phi}\|^2_{\underline{X}}) \le c$. Futhermore, $\bar{X}$ has the following continuity properties,

$$\text{(ii)} \qquad \lim_{t' \to t+0} E\|\bar{X}(t') - \bar{X}(t)\|^2_{\underline{X}} = 0 \quad \forall t \in [t_0,T)$$

$$\text{(iii)} \qquad X(\cdot,a) \in L_2(t_0,T) \text{ and } \lim_{a' \to a} E\|X(\cdot,a') - X(\cdot,a')\|^2_{L_2(t_0,T)} = 0, \quad \forall a > 0.$$

If, in addition to (H), β satisfies (for some $\mu \in (0,1]$ and every $A > 0$) the condition,

$$(\bar{H}) \qquad \begin{cases} \int_0^T (\beta(r,r+\alpha) - \beta(r,r+\alpha))^2(r + \alpha')dr \le K_1(A)|\alpha - \alpha'|^\mu, \\ \qquad\qquad \text{for } -T \le \alpha' < \alpha \le A, \\ \\ \int_{(A-T)^+}^A (\beta(b-\alpha',b) - \beta(b - \alpha,b))^2(b - \alpha')db \le K_2(A)|\alpha - \alpha'|^\mu , \\ \qquad\qquad \text{for } (A - T)^+ \le \alpha' < \alpha \le A , \end{cases}$$

(where we put $\beta(s,a) = 0$ whenever $s < 0$) then the solution $\bar{X}$ possesses a stronger continuity property, namely,

$$\text{(iv)} \qquad \lim_{h \to 0^+} E \sup_{t \le u \le t+h} \|\bar{X}(u) - \bar{X}(t)\|^2_{\underline{X}} = 0, \quad \forall t \in [t_0,T).$$

The proof of the theorem is based on the following,

<u>Lemma 1.1.</u> Let Y be defined by

$$(1.1) \qquad Y(t,a) := \int_{t_0}^t \beta(r,a - t + r)\partial_r W(r,a - t + r)$$

for $t_0 \le t \le T$ and $0<a$. Assume that (H) holds. Then, $Y(t,\cdot) \in L_2^\rho$ ($\forall t \in [t_0,T]$) and $Y(\cdot,a) \in L_2(t_0,T)$ ($\forall a \in R_+$) a.s.; $t \to Y(t,\cdot) \in L_2^\rho$ and $a \to Y(\cdot,a) \in L_2(t_0,T)$

are continuous in quadratic mean. Furthermore, $E\|Y(\cdot,a)\|^2_{L_2(t_0,T)} \to 0$ as $a \to 0$. If in addition condition $(\overline{H})$ holds, then

$$(1.2) \qquad \lim_{h\to 0_+} E \sup_{t\le u\le t+h} \|Y(u,\cdot) - Y(t,\cdot)\|^2_{L_2^\rho} = 0, \quad \forall t > t_0.$$

Note. Since $Y(t_0,a) = 0$ ($\forall a \in \mathbb{R}_+$), (H) implies in particular that $E\|Y(t,\cdot)\|^2_{L_2^\rho} \to 0$ as $t \to t_0$.

Using this lemma the existence of a solution $\overline{X}$ is established by the method of successive approximations.

2. The Markov Property

In order to establish the Markov property of the solution process of (E) we need the following.

Lemma 2.1. Assume that β satisfies condition (H) and that g,k satisfy condition (L). Let $\overline{X}$ and $\overline{Z}$ be solutions of (E) corresponding to the initial conditions $\overline{X}(t_0) = \overline{\phi}$, $\overline{Z}(t_0') = \overline{\psi}$ with $t_0' < t_0$, where $E(\|\overline{\phi}\|^2_{\underline{X}}) < \infty$, $E(\|\overline{\psi}\|^2_{\underline{X}}) < \infty$. Then,

$$(2.1) \qquad \lim_{\substack{t_0 \to t_0'+0 \\ \overline{\phi} \tilde{\to} \overline{\psi}}} E \sup_{t_0\le u\le T} \|\overline{X}(u) - \overline{Z}(u)\|^2_{\underline{X}} = 0$$

where $\overline{\phi} \tilde{\to} \overline{\psi}$ means that $E\|\overline{\phi} - \overline{\psi}\|^2_{\underline{X}} \to 0$.

We now adopt the following conventions. For $s \ge 0$ and $\overline{\phi} \in \underline{X}$, $\overline{X}_{s,\overline{\phi}}$ will denote the ($\underline{X}$-valued) solution process of (E) satisfying the (deterministic) initial condition $\overline{X}(s) = \overline{\phi}$, and $\{M_t^{s,\overline{\phi}} : t \ge s\}$ will denote the completed σ-algebras determined by this solution:

$$M_t^{s,\overline{\phi}} = \overline{\sigma}\{X(u,a);\ s \le u \le t,\ a \ge 0\}.$$

Since by Theorem A the process $t \to \overline{X}_{s,\overline{\phi}}(t,\omega)$ is measurable, $M_t^{s,\overline{\phi}}$-adapted and $L^2(\Omega)$-continuous, it is also progressively measurable ([D],[Me]). We will put $B(\underline{X})$ for the σ-algebra of Borel sets of $\underline{X}$ and define the following "transition function":

$$(2.2) \qquad p(s,\overline{\phi},t,\Gamma) := P\{\overline{X}_{s,\overline{\phi}}(t) \in \Gamma\} \quad \text{for } t \geq s \geq 0,\ \overline{\phi} \in \underline{X},\ \Gamma \in B(\underline{X}).$$

One now obtains:

<u>Theorem B.</u> Assume that β satisifes condition (H) and that g,k satisfy condition (L). The function p defined in (2.2) has the following properties:

(a) for each $t \geq s \geq 0$ and $\Gamma \in B(\underline{X})$, the mapping

$$(s,\overline{\phi}) \to p(s,\overline{\phi},t,\Gamma) \text{ is } B(R) \times B(\underline{X})\text{-measurable;}$$

(b) for each bounded continuous $\Phi : \underline{X} \to \mathbb{R}$ and each $t \in (t_0,T]$ the quantity

$$F(s,\overline{\phi}) := E\Phi(\overline{X}_{s,\overline{\phi}}(t)), \quad s < t, \quad \text{satisfies}$$

$$\lim_{\substack{s \uparrow s^* \\ \|\overline{\phi}-\overline{\phi}^*\| \to 0}} F(s,\overline{\phi}) = F(s^*,\overline{\phi}^*);$$

(c) for each $t_2 > t_1 \geq s$ and $\Gamma \in B(\underline{X})$

$$P\{\overline{X}_{s,\overline{\phi}}(t_2) \in \Gamma | M_{t_1}^{s,\overline{\phi}}\} = P\{\overline{X}_{s,\overline{\phi}}(t_2) \in \Gamma | \overline{X}(t_1)\} = p(s,\overline{X}_{s,\overline{\phi}}(t_1),t_1,\Gamma),$$

where the last term on the right denotes the composite of $p(x,\cdot,t_1,\Gamma)$ with $\overline{X}_{s,\overline{\phi}}(t_1)$, and where the other terms utilize the standard notation for conditional probabilities/expectations (e.g. [IW]).

The proof, similar to the one utilized in the case of stochastic differential equations, is only sketched [cf. [GS,III], [IW,Ch. 4], or [Kr,Ch. 2§9] for the finite-dimensional s.d.e case, and [E,Ch.9§5] for a Hilbert space version (involving, however, the Fiske-Stratonovich integral)].

We note that (b) follows directly from Lemma 1 by the bounded convergence theorem, applied to functions $\Phi(X_{s_n,\phi_n}(t,\omega))$ which, by (2.1), converge in probability. Then (a) follows from (b) by the observation that the indicator function 1_Γ is a monotone limit of a sequence of functions Φ of the type considered in (b) [B, p. 8]. However (c) depends on the uniqueness result in Theorem A, together with the fact that time increments of the one- and two-parameter Brownian motions are independent and satisfy:

$$\text{for each } t_0 > 0, \text{ the processes}$$
$$t \to (w(t_0 + t) - w(t_0)), \quad (t,a) \to (W(t_0 + t,a) - W(t_0,a))$$

are a one- and a two-dimensional Brownian motion, respectively. We omit this argument, which is patterned on the arguments given in the cited references.

Note: The conclusions of Theorem B can be reinterpreted: (a) and (c) mean that the process $\{\overline{X}_{s,\overline{\phi}}(t), \overline{\phi} \in \underline{X}, t > s > 0\}$ is a Markov process in $\underline{X}$; (b) then ensures via Theorem 5.9 in [Dy 1] that this process is a strong Markov process (for time homogeneous processes, (b) is the Feller property).

3. The Infinitesimal Generator

We start by recalling,

Definition. The weak infinitesimal generator associated with (E) is the unbounded linear operator A whose domain of definition D_A is the set of globally bounded continuous functionals $\Phi: \mathbb{R}_+ \times \underline{X} \to \mathbb{R}$ such that

(a) The following limit exists for every $(s,\overline{\phi}) \in R_+ \times \underline{X}$

$$(*) \qquad \lim_{h \downarrow 0} \frac{1}{h} [E\Phi(s + h, \overline{X}_{s,\overline{\phi}}(s+h)) - \Phi(s,\overline{\phi})].$$

(b) There exists $h_0 > 0$ such that the converging family in (*) is bounded for $h \in (0,h_0]$, uniformly w.r. to $(s,\overline{\phi}) \in \mathbb{R}_+ \times \underline{X}$.

For $\Phi \in D_A$, $A\Phi$ is defined as the limit in (*).

If condition (b) is dropped, the corresponding operator will be called the extended weak infinitesimal generator.

We shall denote by $C_b^j(\mathbb{R}_+ \times \underline{X})$ the set of bounded continuous functionals on $\mathbb{R}_+ \times \underline{X}$ which possess bounded continuous derivatives up to order j. Further we shall denote by $C^j_{b,\underline{X}}(\mathbb{R}_+ \times \underline{X})$ the set of bounded continuous functionals on $\mathbb{R}_+ \times \underline{X}$ which possess bounded and continuous derivatives (up to order j) with respect to the $\underline{X}$ variable.

We also denote,

$$\overline{H}_1^\rho = \{(r,\phi) \in \underline{X} \colon \phi \in \mathrm{LAC},\ \ \phi(0) = r \ \text{ and } \ \phi^\bullet \in L^2_\rho\}.$$

Here LAC denotes the space of locally absolutely continuous functions on $[0,\infty)$. Further we denote by $\overset{\circ}{\overline{H}}{}_1^\rho$ the subset of $\overline{H}_1^\rho$ consisting of elements of the form $(0,\phi)$. Finally we denote by $\overline{H}_{-1}^\rho$ the dual space of $\overset{\circ}{\overline{H}}{}_1^\rho$.

The following result describes a subset of D_A on which the extended weak infinitesimal generator A can be represented as a differential operator.

Theorem C

In addition to the hypotheses of Theorem B, suppose that g,k,β are continuous and that β satisfies the conditions,

i) $\int_0^\infty (1 + a^2)\beta^2(r,a)\rho(a)da < \infty \qquad \forall r \in R_+$

ii) $r \to \int_0^\infty (1 + a)\beta^2(r,a)\rho(a)da$ is locally bounded.

(H_c) iii) For every $(s,a) \in \mathbb{R}_+^2$ and every $\varepsilon > 0$ there exist positive numbers $\delta(\varepsilon)$ and $\sigma(s,a)$ s.t.

$$|\beta(s',a') - \beta(s,a)| < \delta(\varepsilon)\sigma(s,a)$$

whenever $|s' - s| + |a' - a| < \varepsilon$, and $\delta(\varepsilon) \to 0$ as $\varepsilon \to 0$ while σ is measurable and

$$\int_0^\infty (1 + a)\sigma^2(s,a)\rho(a)da < M_s \qquad \forall s \in [0,\infty),$$

where M_s is locally bounded.

Assume also that ρ is locally absolutely continuous with $\dot{\rho}/\rho$ essentially bounded. Then the set E defined below is in the domain of the extended weak infinitesimal generator,

$$E := C_b^1(R_+ \times \underline{X}) \cap C^2_{b,\underline{X}}(R_+ \times \underline{X}) \cap C^1_{b,\overline{H}^\rho_{-1}}(R_+ \times \overline{H}^\rho_{-1})$$

and for $\Phi \in E$, $A\Phi$ is given by

$$A\Phi(s,\overline{\phi}) = \Phi_{,t}(s,\overline{\phi}) = g(s,\overline{\phi})\Phi_{,1}(s,\overline{\phi}) \tag{3.1}$$

$$+ \frac{1}{2} k^2(s,\overline{\phi})\Phi_{,11}(s,\overline{\phi}) + \frac{1}{2} tr(\Phi_{,22}(s,\overline{\phi})U_s)$$

$$+ \langle(\Phi_{,2}(s,\overline{\phi}))^{\bullet} + \Phi_{,2}(s,\overline{\phi})\dot{\rho}/\rho, \phi\rangle_{L_2^\rho}$$

for all $(s,\overline{\phi}) \in R_+ \times \underline{X}$. Here $\Phi_{,1}$ (resp. $\Phi_{,2}$) denotes (Frechet) differentiation with respect to the first (resp. second) component of $\overline{\phi} = (\phi_0,\phi(\cdot))$, $\Phi_{,22}$ is interpreted as a bounded linear operator on L_2^ρ and U_s is the Hilbert-Schmidt operator on L_2^ρ with kernel

$$G_s(a,b) = (a \wedge b)\beta(s,a)\beta(s,b).$$

<u>Note.</u> $(\Phi_{,2}(s,\overline{\phi}))^{\bullet}$ denotes the derivative of $\Phi_{,2}(s,\overline{\phi})$ as an element of $\overset{\circ}{H}{}^\rho_1$.

The proof of the theorem is lengthy and rather technical; therefore we shall not discuss it here.

4. Properties of the Infinitesimal Generator

Under appropriate conditions on the diffusion coefficients β,k we can easily verify that the infinitesimal generator A is non-singular.

Assuming that k^2 is positive everywhere in $R^+ \times \underline{X}$ it only remains to verify that the operator $U_s : L_2^\rho(0,\infty) \to L_2^\rho(0,\infty)$ which is represented by the kernel

(4.1) $$G_s(a,b) = \beta(s,a)\beta(s,b)a \wedge b$$

is positive definite for every $s \in R^+$. We shall show that this is the case if $\beta(s,\cdot)$ satisfies the following additional conditions for every $s \in R^+$.

(4.2) The functions $\beta(s,\cdot)$ and $a \to \sqrt{a}\,\beta(s,a)$ are in $L^2_\rho(0,\infty)$ and $\beta(s,\cdot) \neq 0$ a.e. in R^+.

Given $\phi \in L_2^\rho(0,\infty)$ denote,

$$\gamma_s(a) = \int_0^a b\beta(s,b)\phi(b)\rho(b)db,$$

(4.3) $$\delta_s(a) = \int_a^\infty \beta(s,b)\phi(b)\rho(b)db.$$

Each of these is a continuous function on $[0,\infty)$. Now

$$\langle U_s\phi,\phi\rangle_{L^2_\rho} = \int_0^\infty\Big(\int_0^\infty G_s(a,b)\phi(b)\rho(b)db\Big)\phi(a)\rho(a)da$$

$$= \int_0^\infty\int_0^a b\beta(s,a)\beta(s,b)\phi(b)\rho(b)db\,\phi(a)\rho(a)da$$

$$+ \int_0^\infty\int_a^\infty a\beta(s,a)\beta(s,b)\phi(b)\rho(b)db\,\phi(a)\rho(a)da$$

$$= \int_0^\infty \gamma_s(a)\beta(s,a)\phi(a)\rho(a)da + \int_0^\infty f_s(a)a\beta(s,a)\phi(a)\rho(a)da.$$

Thus,

$$\langle U_s\phi,\phi\rangle_{L^2_\rho} = -\int_0^\infty \gamma_s(a)\delta_s'(a)da + \int_0^\infty \delta(a)\gamma_s'(a)da$$

$$= 2\int_0^\infty \delta_s(a)\gamma_s'(a)da.$$

Here we used the fact that $\lim_{a\to\infty} \delta_s(a)\gamma_s(a) = 0$. This is a consequence of (4.2) and the inequalities,

$$(4.4)\qquad \begin{aligned}\delta_s^2(a) &< \int_a^\infty \beta^2(s,b)\rho(b)db\,\|\phi\|^2_{L^2_\rho}\\ &< \frac{1}{a}\int_a^\infty b\beta^2(s,b)\rho(b)db\,\|\phi\|^2_{L^2_\rho}\end{aligned}$$

and

$$\gamma_s(a) < (a\int_0^\infty b\beta^2(s,b)\rho(b)db\int_0^\infty \phi^2(b)\rho(b)db)^{1/2}\ .$$

Next, observing that $\gamma_s'(a) = -a\delta_s'(a)$, we obtain

$$\langle U_s\phi,\phi\rangle_{L^2_\rho} = -2\int_0^\infty a\delta_s'(a)\delta_s(a)da = \int_0^\infty \delta_s^2(a)da$$

since $a\delta_s^2(a) \to 0$ as $a \to \infty$ by (4.4). Finally, differentiating (4.3) and using (4.2), we deduce that δ is identically zero only if $\phi = 0$ a.e.

Under some additional conditions it can also be shown that the operator U_s is nuclear. This is a key ingredient in treating initial value problems for parabolic equations in an infinite dimensional space (see [Da]). Such problems are encountered in the discussion of stability questions of the kind mentioned in the introduction. We mention also that a somewhat similar problem arises in the study of controlled diffusion processes and was treated in [BD]. However the solution obtained in [BD] (which can be presented in explicit form) is a weak solution in the sense that it satisfies the equation only on a dense subset of the infinite dimensional space. Such a weak solution would not be appropriate for the applications that we wish to study.

In some applications it is necessary to deal with initial-boundary value problems rather than pure initial value problems. For instance this is the case when dealing with the concept of expected exit time in relation to stability (see [S1,2]). A boundary value elliptic problem in a Hilbert space has been treated in [F] using the pioneering work of Daletskii [Da]. Frolov's approach applies to equations which are separable in a certain sense. This feature is present in our problem with respect to the main part of the operator but not with respect to the lower order terms. However it seems that the problem can be treated by using the ideas of Daletskii and Frolov complemented by appropriate a-priori estimates. We plan to present a discussion of these problems in a forthcoming paper.

References

[B] P. Billingsley, Convergence of Probability Measures, Wiley, New York, 1968.

[BD] V. Barbu and G. DaPrato, "Existence for the Dynamic programming equations for control diffusion processes in Hilbert space", Sc. Norm. Sup. Pisa (1982).

[CM] B.D. Coleman and V.J. Mizel, "Norms and Semigroups in the theory of fading memory," Arch. Rational Mech. Anal. **23** (1966), 87-123.

B.D. Coleman and V.J. Mizel, "On the general theory of fading memory," Arch. Rational Mech. Anal. **29** (1968), 18-31.

[CW] R. Cairoli and J.B. Walsh, "Stochastic integrals in the plane," Acta Math. **134** (1975), 111-183.

[Da] Yu. L. Daletskii, "Infinite-dimensional elliptic operators and parabolic equations connected with them" Russ. Math. Surveys **22** (1967), 1-53.

[D] J.L. Doob, Stochastic processes, Wiley, New York, 1967.

[Dy] E.B. Dynkin, Theory of Markov Processes, Prentice-Hall, 1961.

E.B. Dynkin, Markov Processes I,II, Academic, Springer, 1965.

[E] K.D. Elworthy, Stochastic differential equations on manifolds, Lond. Math. Soc. Lect. Notes #70, Cambridge Univ. Press, Cambridge, 1982.

[F] N.H. Frolov, "On the Dirichlet problem for an elliptic operator in a cylindrical domain of Hilbert space," Math. USSR Sbornik **21** (1973), 423-438.

[GS] I.I. Gihman and A.V. Skhorohod, Stochastic Differential Equations, Springer, 1972.

I.I. Gihman and A.V. Skhorohod, The Theory of Stochastic Processes III, Grundlehren Math. Wissenschaften #218, Springer, New York, 1975.

[H] R.Z. Hasminskii, Stochastic Stability of Differential Equations, Sijthoff and Noordhoff, 1980.

[IW] N. Ikeda and S. Watanabe, Stochastic Differential Equations and Diffusion Processes, North Holland, 1981.

[Kr] N.V. Krylov, Controlled diffusion processes, Springer, New York, 1980.

[Ku] H.J. Kushner, "On the stability of processes defined by stochastic difference-differential equations," J. Differential Equations **4** (1968), 424-443.

[L] S. Lang, Introduction to differentiable manifolds, Interscience, 1962.

[MM] M. Marcus and V.J. Mizel, "Nonlocal boundary problems for functional partial differential equations," in Integral and Functional Differential Equations, Dekker, New York (1981), 95-108.

M. Marcus and V.J. Mizel, "Semilinear hereditary hyperbolic systems with nonlocal boundary conditions, A&B", J. Math. Anal. and Appls.'s **76** (1980), 440-475; **77** (1980), 1-19.

[Me] P.A. Meyer, Probability and Potential, Blaisdel, Waltham, 1966.

[Mo] S.E.A. Mohammed, "The infinitesimal generator of a stochastic functional differential equation", in Proceedings of Dundee Conference on Ordinary and Partial Differential Equations, 1982, Springer Lecture Notes in Math. #964, Springer, New York, 529-536.

S.E.A. Mohammed, Stochastic Functional Differential Equations, Pitman Research Notes in Math. #99, Pitman, Boston, 1984.

[MT] V. Mizel and V. Trutzer, "Stochastic hereditary equations: existence and asymptotic stability", J. Integral Eqns. 7 (1984), 1-72.

[S] Z. Schuss, "Regularity theorems for solutions of a degenerate evolution equation, Arch. Rat. Mech. Anal. **46** (1972), 200-211.

Z. Schuss, "Backward and degenerate parabolic equations," Applic. Anal. 7 (1978), 111-119.

Z. Schuss, "Singular perturbation methods in stochastic differential equations of mathematical physics," SIAM J. Review **22** (1980), 119-155.

[W] E. Wong, Introduction to Random Processes, Springer, New York, 1983.

[WZ] E. Wong and M. Zakai, "Martingale and stochastic integrals for processes with a multidimensional parameter," Z. Wahrsch. u.Verw. Geb. **29** (1974), 109-122.

DEVELOPMENT OF SINGULARITIES IN NONLINEAR VISCOELASTICITY

J. A. Nohel
Mathematics Research Center
University of Wisconsin-Madison
Madison, WI 53705/USA
and
M. J. Renardy
Department of Mathematics
Virginia Polytechnic Institute and State University
Blacksburg, VA 24061/USA

ABSTRACT. We discuss the motion of nonlinear viscoelastic materials with fading memory in one space dimension. We formulate the mathematical problem, survey results for global existence of classical solutions to the initial value problem if the data are sufficiently small, and discuss in detail the development of singularities in initially smooth solutions for large data.

1. INTRODUCTION AND DISCUSSION OF RESULTS.

In this paper we discuss the motion of nonlinear viscoelastic materials with fading memory in one space dimension. We concentrate on viscoelastic solids and briefly remark on simlar results for fluids. After formulating the mathematical problems, we survey results for global existence of classical solutions to the initial value problem, provided the initial data are sufficiently small. We then discuss in some detail the development of singularities in initially smooth solutions for large data.

We consider the longitudinal motion of a homogeneous one-dimensional body occupying an interval B in a reference configuration and having unit reference density. For simple materials, the stress σ at a material point x is a nonlinear functional of the entire history of the strain $\varepsilon = u_x$ at the same point x (here u denotes the displacement). In this paper, we confine ourselves to the following model problem, which

Sponsored by the United States Army under Contract No. DAAG29-80-C-0041. The work of M. Renardy was supported in part by the National Science Foundation under Grant No. MCS-8215064.

can be motivated as a natural generalization of Boltzmann's constitutive relation for linear viscoelasticity [1] (the derivation of similar results in a variety of other models will be discussed in a later paper)

$$\sigma(x,t) = \varphi(\varepsilon(x,t)) + \int_{-\infty}^{t} a'(t-\tau)\psi(\varepsilon(x,\tau))d\tau , \qquad (x \in B, -\infty < t < \infty) . \tag{1.1}$$

Here φ and ψ are given smooth functions $\mathbf{R} \to \mathbf{R}$ with

$$\varphi(0) = \psi(0) = 0, \ \varphi' > 0, \ \psi' > 0 , \tag{1.2}$$

and for physical reasons the relaxation function $a : [0,\infty) \to \mathbf{R}$ is positive, decreasing, convex, and $a' \in L^1[0,\infty)$, where $'$ denotes the derivative. The conditions on a imply that the stress relaxes as time increases and that deformations which occurred in the distant past have less influence on the present stress than those which occurred more recently. Since only a' occurs in the equation, we may use the normalization $a(\infty) = 0$. In the rheological literature the relaxation function a is often taken to be a finite linear combination of decaying exponentials with positive coefficients obtained by a least square fit to experimental data.

When (1.1) is substituted into the balance of linear momentum, the following integrodifferential equation for the displacement u results:

$$u_{tt} = \varphi(u_x)_x + a'*\psi(u_x)_x + f , \quad x \in B, \ t > 0 . \tag{1.3}$$

Here $*$ denotes the convolution $(\alpha*\beta)(t) = \int_0^t a(t-\tau)\beta(\tau)d\tau$, and f is the sum of an external body force and the history term $\int_{-\infty}^{0} a'(t-\tau)\psi(u_x(x,\tau))_x d\tau$. An appropriate dynamical problem is to determine a smooth function $u : B \times (0,\infty) \to R$ which satisfies (1.3) together with appropriate boundary conditions if B is bounded, and which at $t = 0$ satisfies prescribed initial conditions

$$u(x,0) = u_0(x), \ u_t(x,0) = u_1(x), \ x \in B$$

for certain smooth functions u_0 and u_1. To avoid technical complications we assume in the following that $f = 0$. We also restrict ourselves to the case of an unbounded body, $B = \mathbf{R}$ and we study the Cauchy problem

$$u_{tt} = \varphi(u_x)_x + a'*\psi(u_x)_x , \quad x \in \mathbf{R}, \ t > 0 , \tag{1.4}$$

$$u(x,0) = u_0(x),\ u_t(x,0) = u_1(x) \quad , \quad x \in R \quad . \qquad (1.5)$$

When $a' \equiv 0$ and φ satisfies (1.2), the body is purely elastic. In this case it is well known (see Lax [14], MacCamy and Mizel [15], Klainerman and Majda [13]), that in general the Cauchy problem (1.4), (1.5) does not have globally defined smooth solutions, no matter how smooth and small the initial data are chosen. The initially smooth solution u develops singularities (shock waves) in finite time.

If $a' \not\equiv 0$ and a satisfies the sign conditions above, the fading memory term in (1.4) introduces a weak dissipation mechanism. Significant insight into the strength of this mechanism was gained by the work of Coleman and Gurtin [2], who studied the growth and decay of acceleration waves in materials with memory. They showed that the amplitude $q(t)$ of an acceleration wave propagating into a homogeneously strained medium at rest satisfies a Bernoulli-Riccati ordinary differential equation. The coefficient of q^2 in this equation is proportional to a second order elastic modulus, which is given by φ'' in our model problem, and there is a linear damping term proportional to $a'(0)$. Thus the amplitude $q(t) = [u_{tt}]$ decays to zero as $t \to \infty$, provided $|q(0)|$ is sufficiently small. On the other hand, if $\varphi'' \neq 0$, then $q(t) \to \infty$ in finite time if $|q(0)|$ is large enough, and $q(0)$ is of a certain sign. They did not study existence of such solutions.

This suggests that, under appropriate assumptions on φ, ψ and a, the Cauchy problem (1.4), (1.5) should have unique, globally defined classical (C^2) solutions for sufficiently smooth and small initial data u_0, u_1, while smooth solutions should develop singularities in finite time if the initial data are large in an appropriate sense. Such a global existence result for small data was recently established by Hrusa and Nohel [10] using delicate a priori estimates obtained by combining an energy method with properties of Volterra equations (even in the presence of a small body force). We refer to a recent survey [9] for earlier small data results on initial boundary value problems modelling the motion of finite viscoelastic bodies, and for technical simplifications of the analysis in the special cases $\varphi \equiv \psi$ or $a(t) = e^{-t}$. For the global results the Cauchy problem is more difficult than the finite body problem because the

Poincaré inequality is not available to estimate lower order derivatives from higher order derivatives.

The remainder of our discussion will deal with the formation of singularities in finite time from smooth solutions of the Cauchy problem (1.4), (1.5). For the special case $\varphi \equiv \psi$, Markowich and Renardy [17] have obtained numerical evidence for the formation of shock fronts in finite time from large data, and Hattori [7] has shown that, if $\varphi'' \not\equiv 0$ and if the body B is finite, then there are smooth initial data (which he does not characterize) for which the corresponding Dirichlet-initial value problem does not have a globally defined smooth solution. On the other hand, Hrusa [8] has shown that if φ is linear and only ψ is allowed to be nonlinear, then the Cauchy problem (1.4), (1.5) does have globally smooth solutions, even for large smooth data. Therefore, we shall restrict ourselves to the case when $\varphi'' \neq 0$, at least over the range of the solution. The case when φ'' changes sign will require further refinements.

An essential ingredient in the analysis (which is also important for the global theory) is the following local existence result which is established by combining Banach's fixed point theorem on an appropriate function space with standard energy estimates and Sobolev's embedding theorem.

Proposition 1:
Assume that $\varphi, \psi \in C^3(\mathbf{R})$ satisfy (1.2); assume $a, a', a'' \in L^1_{\ell oc}[0,\infty)$,(*) and there is a constant $\kappa > 0$ such that

$$\varphi'(\xi) \geq \kappa, \quad \xi \in \mathbf{R} .$$

Assume that $u_0 \in L^2_{\ell oc}(\mathbf{R})$ and that $u_0', u_1 \in H^2(\mathbf{R})$. Then the Cauchy problem (1.4), (1.5) has a unique classical solution $u \in C^2(\mathbf{R} \times [0,T_0))$ defined on a maximal interval $(0,T_0)$. If T_0 is finite, then

$$\sup_{\mathbf{R}\times[0,T_0)} [|u_{xx}(x,t)| + |u_{xt}(x,t)|] = \infty .$$

The proof of Proposition 1 is almost identical to that of Theorem 2.1 of [6], and we omit the details; only certain readily available energy estimates for lower order derivatives are

(*) Here the square bracket means integrability up to 0. No sign condition on a are required, but $a'(0)$ finite is essential.

needed. The characterization of the maximal interval of existence is established by combining the energy estimates obtained in [6] with a Gronwall inequality argument. We remark that the energy estimates used in the proof of Proposition 1 yield time-dependent bounds which cannot be used to obtain global estimates. These can only be constructed by taking advantage of the damping mechanism induced by the memory term under appropriate sign conditions on a and by assuming the initial data to be small (see [10] for details).

The assumptions concerning the kernel a in Proposition 1 imply that a' is absolutely continuous on $[0,\infty)$. Recently, Hrusa and Renardy [11] established a result similar to Proposition 1 (and proved a global existence result for small data for bounded bodies) under assumptions which permit a singularity in a' at $t = 0$ (e.g. $a'(t) \sim -t^{\alpha-1}$, $0 < \alpha < 1$ as $t \to 0^+$). Such singularities are relevant for certain popular models of viscoelastic materials.

Our main result on development of singularities for large enough data is

Theorem 1:

Let $\varphi, \psi \in C^3(\mathbf{R})$ satisfy (1.2) and assume $a, a', a'' \in L^1_{\ell oc}[0,\infty)$. Assume that $\varphi''(0) \neq 0$. Then, for every $T_1 > 0$, we can choose initial data $u_0', u_1 \in C^2(\mathbf{R}) \cap L^\infty(\mathbf{R})$ such that the maximal time interval of existence, given by Proposition 1, for the smooth solution of the Cauchy problem (1.4), (1.5) cannot exceed T_1. More precisely, if $\sup_{x\in\mathbf{R}} |u_0'(x)|$ and $\sup_{x\in\mathbf{R}} |u_1(x)|$ are sufficiently small, while $u_0''(x)$ and $u_1'(x)$ assume sufficiently large values with appropriate signs, then there is some $t^* < T_1$ such that

$$\sup_{\mathbf{R}\times[0,t^*)} \{|u_{xx}(x,t)| + |u_{xt}(x,t)|\} = \infty , \qquad (1.6)$$

while

$$\sup_{\mathbf{R}\times[0,t^*)} \{|u_x(x,t)| + |u_t(x,t)|\} < \infty \qquad (1.7)$$

(and in fact, this latter quantity remains small).

In view of the analogy with hyperbolic conservation laws and the numerical evidence [17], it is to be expected that a blow-up as established by Theorem 1 will lead to the development of a shock front.

The method of the proof, sketched in section 2 is to show that the memory term is in fact of lower order than the elastic term $\varphi(u_x)_x$ and can be treated as a perturbation. While considerably more technical, the proof is a generalization of the approach of Lax [14] for showing the development of singularities for the quasilinear wave equation

$$u_{tt} = \varphi(u_x)_x \quad .$$

Theorem 1 was established independently by Dafermos [4] using an approach which is different from ours but similar in spirit. The result can also be established by modifying the results of F. John [12] and extending them to systems of quasilinear hyperbolic conservation laws which contain lower order source terms (F. John, private communications).

Similar results for first order model problems were derived by Malek-Madani and Nohel [16] and, using different methods, by Renardy [18] and Dafermos [3].

A particular case of the model equation studied in this paper leads to a model for shearing flows of viscoelastic fluids studied recently by Slemrod [20]. With $v(x,t)$ denoting the velocity of the fluid in simple shear, Slemrod studies the problem

$$\begin{aligned} v_t &= a*\varphi(v_x)_x \quad , \quad (x \in R,\ t > 0) \quad , \\ v(x,0) &= v_0(x) \quad , \quad (x \in R) \quad . \end{aligned} \tag{1.8}$$

for the special case $a = e^{-t}$. Problem (1.8) leads to a Cauchy problem of the form (1.4), (1.5) after differentiation with respect to time. Then Theorem 1 can be used to get a blow-up result for this problem, like the result found by Slemrod for $a(t) = e^{-t}$. The global existence of solutions for small data follows from [5, Theorem 4.1]. Other popular models for viscoelastic fluids have been analyzed by the method used in this paper; the results will be published elsewhere (see Note at end of paper).

It is an open problem whether Theorem 1 holds for kernels a in (1.4) with a' having an integrable singularity at zero.

2. **Development of Shocks.**

In this section, we sketch the proof of Theorem 1 establishing the development of shocks from initially smooth solutions of the Cauchy problem (1.4), (1.5) in finite time. For simplicity, most of the analysis will be carried out for the special case $a(t) = e^{-t}$; the proof for more general relaxation functions as well as for a more general class of model equations will be carried out in a forthcoming paper.

We begin by transforming (1.4) to an equivalent system. We let $w = u_x$, $v = u_t$, and write the constitutive assumption (1.1) in the form

$$\sigma = \varphi(w) - z \quad , \quad z = -a'*\psi(w) \quad .$$

Since we have assumed $\varphi' > 0$, the first of these equations can be solved for w,

$$w = \varphi^{-1}(\sigma+z) =: g(\sigma,z) \quad ,$$

and g is a smooth function of $\sigma \in \mathbf{R}$, $z \in \mathbf{R}$. As long as the solution remains smooth, the Cauchy problem (1.4), (1.5) is equivalent to the first order system

$$\begin{aligned} v_t &= \sigma_x \;, \\ \sigma_t &= c^2(\sigma,z)v_x + a'(0)\psi(g(\sigma,z)) + a''*\psi(g(\sigma,z)) \;, \\ z_t &= -a'(0)\psi(g(\sigma,z)) - a''*\psi(g(\sigma,z)) \quad . \end{aligned} \tag{2.1}$$

The initial conditions become

$$v(x,0) = u_1(x), \; \sigma(x,0) = \varphi(u_0'(x)), \; z(x,0) = 0 \quad . \tag{2.2}$$

By c we have denoted the wave speed

$$c(\sigma,z) := [\varphi'(g(\sigma,z))]^{1/2} \quad ;$$

c is a smooth function of σ and z. The system (2.1) is hyperbolic, and its eigenvalues are $+c$, $-c$ and 0. Under the assumptions of Proposition 1, a C^1-solution exists on some maximal interval $\mathbf{R} \times [0,T_0)$. If T_0 is finite, then the first derivatives of v, σ, z must become infinite as $t \to T_0$. It is immediate from equation (2.1) that σ_t, σ_x, z_t and z_x will remain bounded as long as v, σ, z, v_t and v_x are bounded.

To proceed further, we extend the classical approach of Lax [14] for first order hyperbolic 2 × 2-systems. We define "approximate" Riemann invariants by those quantities which would be the classical Riemann invariants if z in the first two equations of (2.1) were treated as a parameter. These quantities are given by

$$\begin{aligned} r &= r(v,\sigma,z) = v + \Phi(\sigma,z) \quad , \\ s &= s(v,\sigma,z) = v - \Phi(\sigma,z) \quad , \\ \Phi(\sigma,z) &= \int_{\sigma_0}^{\sigma} \frac{d\zeta}{c(\zeta,z)} \quad ; \end{aligned} \tag{2.3}$$

without loss of generality we may take $\sigma_0 = 0$. Since $\Phi_\sigma(\sigma,z) = \frac{1}{c(\sigma,z)} > 0$, this correspondence is smoothly invertible, and we have

$$v = \frac{r+s}{2} \ , \ \Phi(\sigma,z) = \frac{r-s}{2} \quad .$$

In the following, we assume $a(t) = e^{-t}$. Then (2.1) takes the simple form

$$\begin{aligned} v_t &= \sigma_x \quad , \\ \sigma_t &= c^2(\sigma,z)v_x - \psi(g(\sigma,z)) + z \quad , \\ z_t &= \psi(g(\sigma,z)) - z \quad . \end{aligned} \tag{2.4}$$

We now differentiate r and s along the c and $-c$ characteristics, respectively, and z along the zero characteristic (i.e. we form $r_t - cr_x$, $s_t + cs_x$ and z_t). This leads to the following first order hyperbolic system equivalent to (2.4), (2.2)

$$\begin{aligned} r_t - Ar_x &= -Bz_x + CD \quad , \\ s_t + As_x &= -Bz_x - CD \ , \\ z_t &= D \quad , \end{aligned} \tag{2.5}$$

with the initial data

$$\begin{aligned} r(x,0) &= u_1(x) + \Phi(\varphi(u_0'(x)),0) \quad , \\ s(x,0) &= u_1(x) - \Phi(\varphi(u_0'(x)),0) \quad , \\ z(x,0) &= 0 \quad ; \end{aligned} \tag{2.6}$$

$$A = A(r,s,z) := c(\sigma(r,s,z),z) > 0 \quad ,$$

$$B = B(r,s,z) := c(\sigma(r,s,z),z)\Phi_z(\sigma(r,s,z),z) \quad ,$$

$$C = C(r,s,z) := \Phi_z(\sigma(r,s,z),z) - \frac{1}{c(\sigma(r,s,z),z)} \quad ,$$

$$D = D(r,s,z) := \psi(g(\sigma(r,s,z),z)) - z \quad .$$

To establish the development of shocks in finite time, we study the evolution along characteristics of the quantities

$$\begin{aligned} \rho &:= v_x + \frac{\sigma_x}{c(\sigma,z)} \quad , \\ \tau &:= v_x - \frac{\sigma_x}{c(\sigma,z)} \quad , \end{aligned} \tag{2.8}$$

and z_x. Note that if z were a constant parameter, then ρ and τ would be the x-derivatives of r and s. We have $v_x = \frac{1}{2}(\rho+\tau)$, $\sigma_x = \frac{1}{2}c(\rho-\tau)$, and

$$(c^2)_\sigma(\sigma,z) = 2cc_\sigma = \frac{\varphi''(g(\sigma,z))}{\varphi'(g(\sigma,z))}.$$

A tedious but straightforward calculation using the relations (obtained by differentiating (2.4))

$$\begin{aligned} v_{tx} &= \sigma_{xx} \quad , \\ \sigma_{tx} &= c^2(\sigma,z)v_{xx} + (c^2)_\sigma(\sigma,z)\sigma_x v_x \\ &\quad + (c^2)_z(\sigma,z)z_x v_x - D_x \quad , \\ z_{tx} &= D_x \quad , \end{aligned} \tag{2.9}$$

yields the system

$$\begin{aligned} \rho_t - c\rho_x &= \frac{(c^2)_\sigma}{4}\rho(\rho-\tau) + O(|\rho||z_x| + |\tau||z_x| \\ &\qquad + |\rho| + |\tau| + |z_x|) \quad , \\ \tau_t + c\tau_x &= -\frac{(c^2)_\sigma}{4}\tau(\rho-\tau) + O(|\rho||z_x| + |\tau||z_x| \\ &\qquad + |\rho| + |\tau| + |z_x|) \quad , \\ z_{xt} &= O(|\rho| + |\tau| + |z_x|) \quad . \end{aligned} \tag{2.10}$$

subject to the initial data

$$\rho(x,0) = u_1'(x) + \varphi'(u_0'(x))^{1/2} u_0''(x) \;,$$

$$\tau(x,0) = u_1'(x) - \varphi'(u_0'(x))^{1/2} u_0''(x) \;, \qquad (2.11)$$

$$z_x(x,0) = 0 \;.$$

The cross product terms $\rho\tau$ in (2.10) are eliminated if one considers the characteristic derivatives of $c(\sigma,z)^{1/2}\rho$ and $c(\sigma,z)^{1/2}\tau$ (see Lax [14] and Slemrod [19]). We find

$$\left(\frac{\partial}{\partial t} - c\,\frac{\partial}{\partial x}\right)(c^{1/2}\rho) = \gamma(c^{1/2}\rho)^2 + O(|\rho||z_x| + |\tau||z_x| + |\rho| + |\tau| + |z_x|) \;,$$

$$\left(\frac{\partial}{\partial t} + c\,\frac{\partial}{\partial x}\right)(c^{1/2}\tau) = \gamma(c^{1/2}\tau)^2 + O(|\rho||z_x| + |\tau||z_x| + |\rho| + |\tau| + |z_x|) \;, \qquad (2.12)$$

$$z_{xt} = O(|\rho| + |\tau| + |z_x|) \;.$$

Here the coefficient function γ is given by

$$\gamma = \gamma(\sigma,z) = \frac{1}{4}\,\frac{\varphi''(g(\sigma,z))}{\varphi'(g(\sigma,z))^{5/4}} \;.$$

In (2.12) terms like $O(|\rho||z_x|)$ represent a bilinear expression in ρ and z_x with coefficients depending only on r, s, z. For definiteness, let us assume $\varphi''(0) > 0$ (the discussion for $\varphi''(0) < 0$ is analogous). We take initial data with the following properties: u_0' and u_1 (and hence $\rho(x,0)$, $\tau(x,0)$ as well as $z(x,0) \equiv 0$) are uniformly small, and $\rho(x,0)$, $\tau(x,0)$ are such that at least one of them has a large positive maximum (by choosing u_0'' or u_1' or both sufficiently large). At the same time, the maxima of $-\rho$ and $-\tau$ should not be too large.

As long as (r,s,z) remains within a given neighborhood U of 0, we have upper and lower bounds for the coefficients occuring in (2.12); in particular, we have a positive lower bound γ_0 for γ. We shall see later that (r,s,z) will remain in U up to the time of blow-up if they are small enough initially and if we make the maximum of $\rho(x,0)$ or $\tau(x,0)$ large enough.

For every $t > 0$, we now set

$$h(t) = \max[\max_x \rho(x,t),\ \max_x \tau(x,t)] \;.$$

From (2.12), we find that, as long as $(r,s,z) \in U$, while

$h(t)$ is large and $\max_x|z_x| \ll h(t)$, we have, for some positive constants γ_0 and κ

$$(\frac{d}{dt})_+ h(t) > \gamma_0(h(t))^2, \quad \text{and} \quad \max_x|z_{xt}| < \kappa h(t) \ll (h(t))^2 .$$

Initially, we have $|z_x| = 0$ and it follows from these inequalities that it will remain small compared to $h(t)$. We also find that $h(t)$ becomes infinite in finite time. Since there is also some constant γ_1 such that $(\frac{d}{dt})_+ h(t) < \gamma_1(h(t))^2$, it can be shown that, with t^* denoting the blow-up time of h, we have $\frac{c_1}{t^*-t} < h(t) < \frac{c_2}{t^*-t}$ for some constants c_1 and c_2. The third equation of (2.12) then implies that $|z_x|$ grows at most logarithmically as $t \to t^*$. Since $\log(t^*-t)$ is integrable, equations (2.5) imply that r, s, and z remain bounded and in fact small if their initial data are small, and t^* is small (which is the case if $h(0)$ is large). In this way, we can choose the data such that (r,s,z) will in fact remain in U up to the time of blow-up. This completes the sketch of the proof.

References:

[1] L. Boltzmann, Zur Theorie der elastischen Nachwirkung, Ann. Phys. 7 (1876), Ergänzungsband, 624-654.

[2] B. D. Coleman, M. E. Gurtin and I. R. Herrera, Waves in materials with memory, Arch. Rat. Mech. Anal. 19 (1965), 1-19; B. D. Coleman and M. E. Gurtin, ibid, 239-265.

[3] C. M. Dafermos, Dissipation in materials with memory, in: J. A. Nohel, M. Renardy and A. S. Lodge (eds.), Viscoelasticity and Rheology, Academic Press, New York (1985), 221-234.

[4] C. M. Dafermos, Development of singularities in the motion of materials with fading memory, Arch. Rat. Mech. Anal. 91 (1986), 193-205.

[5] C. M. Dafermos and J. A. Nohel, Energy methods for nonlinear, hyperbolic Volterra integrodifferential equations, Comm. PDE 4 (1979), 219-278.

[6] C. M. Dafermos and J. A. Nohel, A nonlinear hyperbolic Volterra equation in viscoelasticity, Amer. J. Math., Supplement (1981), 87-116.

[7] H. Hattori, Breakdown of smooth solutions in dissipative nonlinear hyperbolic equations, Q. Appl. Math. 40 (1982/83), 113-127.

[8] W. J. Hrusa, Global existence and asymptotic stability for a semilinear hyperbolic Volterra equation with large initial data, SIAM J. Math. Anal. 16 (1985), 110-134.

[9] W. J. Hrusa and J. A. Nohel, Global existence and asymptotics in one-dimensional nonlinear viscoelasticity, in: P. G. Ciarlet and M. Roseau (eds.), Trends and Applications of Pure Mathematics to Mechanics, Springer Lecture Notes in Physics 195 (1984), 165-187.

[10] W. J. Hrusa and J. A. Nohel, The Cauchy problem in one-dimensional nonlinear viscoelasticity, J. Diff. Eq. 59 (1985), 388-412.

[11] W. J. Hrusa and M. Renardy, On a class of quasilinear partial integro-differential equations with singular kernels, J. Diff. Eq. 64 (1986), 195-220.

[12] F. John, Formation of singularities in one-dimensional nonlinear wave propagation, Comm. Pure Appl. Math. 27 (1974), 377-405.

[13] S. Klainerman and A. Majda, Formation of singularities for wave equations including the nonlinear vibrating string, Comm. Pure Appl. Math. 33 (1980), 241-263.

[14] P. D. Lax, Development of singularities of solutions of nonlinear hyperbolic partial differential equations, J. Math. Phys. 5 (1964), 611-613.

[15] R. C. MacCamy, A model for one-dimensional nonlinear viscoelasticity, Q. Appl. Math. 35 (1977), 21-33.

[16] R. Malek-Madani and J. A. Nohel, Formation of singularities for a conservation law with memory, SIAM J. Math. Anal. 16 (1985), 530-540.

[17] P. A. Markowich and M. Renardy, Lax-Wendroff methods for hyperbolic history value problems, SIAM J. Num. Anal. 21 (1984), 24-51; Corrigendum, SIAM J. Num. Anal. 22 (1985), 204.

[18] M. Renardy, Recent developments and open prolems in the mathematical theory of viscoelasticity, in: J. A. Nohel, M. Renardy and A. S. Lodge (eds.), Viscoelasticity and Rheology, Academic Press, New York (1985), 345-360.

[19] M. Slemrod, Instability of steady shearing flows in a nonlinear viscoelastic fluids, Arch. Rat. Mech. Anal. 68 (1978), 211-225.

[20] M. Slemrod, Appendix: Breakdown of smooth shearing flow in visco-elastic fluids for two constitutive relations: the vortex sheet vs. the vortex shock, in: D. D. Joseph, Hyperbolic phenomena in the flow of viscoelastic fluids, in: J. A. Nohel, M. Renardy and A. S. Lodge (eds.), Viscoelasticity and Rheology, Academic Press, New York (1985), 309-321.

Note

Added in Proof, February, 1987. Since the completion of this paper in May, 1985, these techniques have been used to obtain finite time blow-up for a variety of one-dimensional models for viscoelastic materials stemming from differential and integral constitutive assumptions, when the data are chosen sufficiently large; see M. Renardy, W. J. Hrusa and J. A. Nohel, Mathematical Problems in Viscoelasticity, Longman Group Limited (formally "Pitman II" monograph series), approximately 275 pages, Chapter II, Section 4 (to appear).

For the Cauchy problem (1.4), (1.5) with a' smooth on $[0,\infty)$ and φ having precisely one inflection point, it has recently been shown in the special case $\psi \equiv \varphi$ that there exists a weak solution (in the class of locally bounded measurable functions) corresponding to arbitrary initial data in $L^\infty(\mathbf{R}) \cap L^2(\mathbf{R})$ by using the method of vanishing viscosity and techniques of compensated compactness (Weak Solutions for a Nonlinear System in Viscoelasticity, J. A. Nohel, R. C. Rogers and A. Tzavaras, in preparation).

MACROMOLECULES IN ELONGATIONAL FLOWS: METASTABILITY AND HYSTERESIS

Yitzhak Rabin*

Center for Studies of Nonlinear Dynamics**
La Jolla Institute
3252 Holiday Court, Suite 208
La Jolla, California 92037

Theoretical [1-7] and experimental [8-9] investigations indicate that dilute solutions of polymers subjected to strong elongational flows undergo a transition from a state in which the macromolecules are in a relatively compact random coil configuration to one in which they are almost fully stretched. While in principle, this coil stretching transition is similar to any other phase transition (in the limit of infinite molecular weight of the polymer), it cannot be treated by the methods of equilibrium statistical mechanics [10], since a salient feature of all the experiments [8,9] is that the time spent by the polymers in the elongational flow field is comparable to their characteristic relaxation time. Thus, one is faced with an intrinsically non-equilibrium situation and in order to analyze the kinetics of the coil stretching transition, one has to rely on methods that were developed in the general context of metastability and nucleation (for systems undergoing first-order phase transitions) [11].

The notion of metastability implies the existence of a time scale T which characterizes the decay of the metastable state. For times much longer than T, thermal fluctuations will populate all the possible states of the system and give rise to a thermal steady-state distribution. In fact, as has been recently demonstrated for a class of dumbbell models with configuration-dependent friction coefficients [7], such a steady-state solution is unique and leads to an average chain extension which is a single-valued function of the applied strain rate. This is a particular example of a more general phenomenon - if one is doing infinite time (>> T) experiments, no hysteresis will be observed in any physical system undergoing a phase transition between its equilibrium states.

---

* Permanent Address: The Weizmann Institute, Rehovot, Israel
** Affiliated with the University of California, San Diego

The situation is quite different if one examines the behavior of the system at finite times ($< T$). Let's assume that a polymer coil is introduced to the flow field (elongational flow with strain rate κ) at time $t = 0$ and spends a time $t = T_{transit}$ in the elongational flow region (Figure 1). The free energy of the polymer in the flow field as a function of its average extension is given in Figure 2. For $t < 0$ there is a single coil minimum C_0; for $0 < t < T_{transit}$ there are two minima, C (coil) and S (stretched state), and a maximum, B (barrier) If rapid switching on (approximated by a step function in Figure 1) of the elongational flow is assumed, the state of the polymer will change from C_0 to C (at $t = 0^+$); i.e. the distribution function will remain peaked about the local free energy minimum corresponding to the coil state. Due to thermal fluctuations (Brownian motion) a double-peaked steady state distribution will eventually be achieved (in the limit of an infinite polymer chain, only the lower minimum corresponding to the stretched state will be populated as $t \to \infty$).

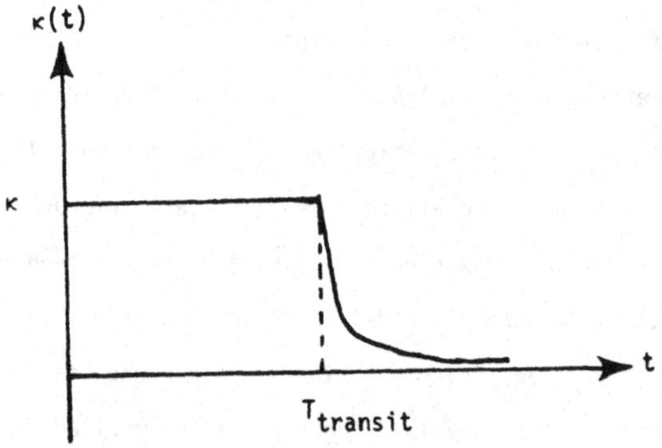

Figure 1

The elongational flow rate $\kappa(t)$ acting on a polymer is plotted as a function of time t.

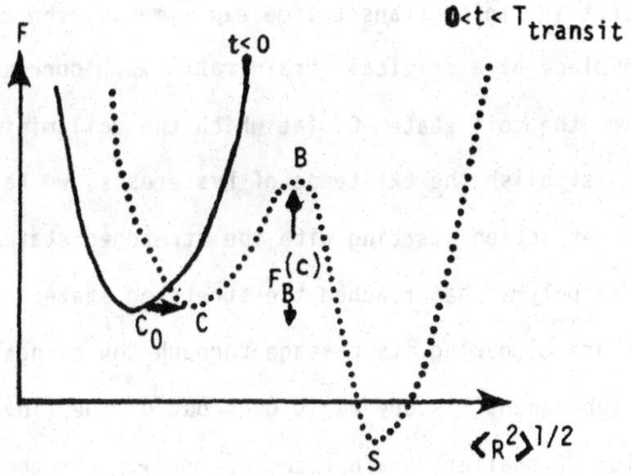

Figure 2

The free energy F of the polymer is plotted against its average elongation $\langle R^2\rangle^{1/2}$, both prior $(t < 0)$ and during $(0 < t < T_{transit})$ its passage through the elongational flow region. The barrier free energy $F_B^{(C)}$ is measured with respect to the coil state C.

The time T_{CS} needed for the establishment of the steady-state via the decay of the metastable coil state C can be estimated using the Becker-Döring theory of nucleation [12]. Taking the inverse of the relaxation time τ_C for a polymer coil as a measure of the rate of fluctuations in the metastable state and $F_B^{(c)}$ as the height of the free energy barrier, we get [11,12]

$$T_{CS} \simeq \tau_C \, e^{F_B^{(c)}/\kappa T} . \quad (1)$$

This time should be compared with the transit time $T_{transit}$ of the polymer in the elongational flow field. Experimental estimates [9] indicate that $T_{transit}$ is of order τ_C and hence that the coil stretching transition can take place during the finite transit time only if the exponential factor in Equation (1) becomes of order unity, i.e., only at strain rates such that the barrier height becomes comparable to κT ($F_B^{(c)}$ is a decreasing function of κ). This is equivalent to saying that in finite transit time experiments, the coil stretching transition will take place at a critical strain rate κ_{CS} corresponding to the stability limit of the coil state C (at which the coil minimum disappears [13]).

In order to establish the existence of hysteresis, we have to consider the process of chain retraction starting with the stretched state of the polymer. Let us assume that the polymer has reached the stretched state (distribution peaked about S_0 in Figure 3) during its passage through the elongational flow region and follows its subsequent history as it gets out of the flow. As the polymer reaches the region of smaller strain rates ($\kappa \ll \kappa_{CS}$ in the exit channel), the coil minimum appears and eventually becomes lower than that corresponding to the stretched state. Repeating the argument of the previous paragraphs, one finds that the time T_{SC} associated with the decay of the metastable stretched state is

$$T_{SC} \simeq \tau_S e^{F_B^{(S)}/\kappa T} \tag{2}$$

where τ_S is the relaxation time for the stretched polymer and $F_B^{(S)}$ is the height of the corresponding free energy barrier (Figure 3).

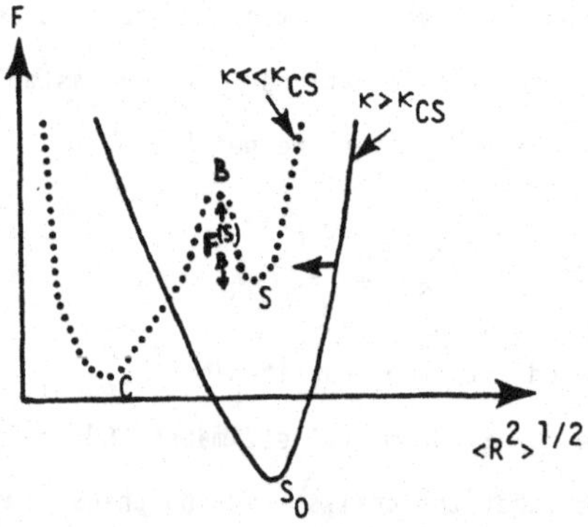

Figure 3

The free energy F of the polymer is plotted against its average elongation $\langle R^2\rangle^{1/2}$. As the polymer gets out of the elongational flow region the strain rate acting on it changes from $\kappa > \kappa_{CS}$ to $\kappa \ll \kappa_{SC}$. The barrier free energy $F_B^{(S)}$ is measured with respect to the stretched state S.

This time should be compared with the finite transit time T_{exit} of the stretched polymer in the exit channel. The relevant time scales in the experiments on chain retraction [8] are $T_{exit} \simeq 2 \cdot 10^{-2}$ sec and $\tau_S \simeq 0.35$ sec, i.e., $T_{exit} << \tau_S$. Thus, the transition to the coil state can take place only when the barrier $(F_B^{(S)})$ becomes of order κT, i.e., at the statibility limit of the stretched state (at a strain rate κ_{SC}).

The critical strain rates κ_{CS} and K_{SC} for the coil stretching and stretched state to coil transitions, respto the coil state can take place only case of ν solvents [6] and, in particular, were shown to obey the scaling laws (up to logarithmic corrections)

$$\kappa_{CS} \sim N^{-3/2} \qquad \kappa_{SC} \sim N^{-2}, \tag{3}$$

where N is the number of statistically independent segments in the polymer. Thus, for long polymers $(N^{1/2} >> 1)$

$$\kappa_{CS}/\kappa_{SC} \sim N^{1/2} >> 1, \tag{4}$$

indicating that much larger strain rates are needed to stretch the polymer from its coil state than to maintain the stretched state once it has been reached.

One can still, of course, ask about the critical strain rate κ_C. This strain rate can be shown [2] to scale as

$$\kappa_C \sim N^{-2}, \tag{5}$$

establishing the hierarchy

$$\kappa_{SC} < \kappa_C << \kappa_{CS}, \tag{6}$$

for sufficiently long chains (this result can be also obtained by application of the lever rule [10] to Figures 5 and 6 in Reference [6]).

The above analysis indicates that while high strain rates $(\kappa_{CS} \sim 10^3 - 10^4 sec^{-1})$ are needed to stretch high molecular weight $(N \sim 10^4)$ polymers in transient elongational flows, much smaller rates $(\kappa_C \sim 10 - 100 sec^{-1})$ are needed to induce the coil stretching transition in steady-state type experiments. This may be of considerable importance for the interpretation of non-Newtonian flow experiments on solutions of high molecular weight polymers, [14] since stretched polymers will have a much larger effect on the flow than coiled ones. The relative effect on the flow can be obtained by comparing the rates of viscous dissipation P_S and P_C in the stretched and coil states respectively. One can show [7] that

$$P_S/P_C \sim N \tag{7}$$

and hence, if even a minute fraction $(1/N)$ of the available polymer undergoes a transition (due to thermal fluctuations) to a stretched state in a steady laminar flow (with an elongational component corresponding to a strain rate $\kappa > \kappa_C$), the effect on the flow can be comparable to that of a much larger (by a factor of N) number of polymer coils.

Acknowledgements

I would like to acknowledge helpful discussions with G. Fuller, R. Bird, M. Renardy, and F. Henyey. This was supported by Defense Advanced Research Projects Agency Contract No. MDA 903-84-C-0373.

References

1. A. Peterlin, J. Chem. Phys. **33**, 1799 (1960).
2. P.G. DeGennes, J. Chem. Phys. **60**, 5030 (1974).
3. R.I. Tanner, Trans. Soc. Rheol. **19**, 556 (1975).
4. E.J. Hinch, Phys. Fluids **20**, No. 10, Pt. 2, S22 (1977).
5. L.G. Leal, G.G. Fuller and W.L. Olbricht, Prog. in Astronautics and Aeronautics **72**, 351 (1980).
6. F.S. Henyey and Y. Rabin, J. Chem. Phys., in press.
7. Xi-jun Fan, R.B. Bird and M. Renardy, J. Non-Newtonian Fluid Mech., in press.

8. C.J. Farrell, A. Keller, M.J. Miles and D.P. Pope, Polymer **21**, 1292 (1980); M.J. Miles and A. Keller, ibid., 1295; A. Keller and J.A. Odell, Colloid and Polymer Sci., in press.

9. G.G. Fuller and L.G. Leal, Rheol. Acta **19**, 580 (1980).

10. L.D. Landau and E.M. Lifshitz, Statistical Physics (Addison-Wesley, Reading, Mass., 1958).

11. J.D. Gunton, M. San Miguel and P.S. Sahni, The Dynamics of First Order Phase Transitions in Phase Transitions and Critical Phenomena, Vol. 8, C. Domb and J.L. Lebowitz, Eds. (Academic Press, London, 1983).

12. R. Becker and W. Döring, Ann Phys. **24**, 719 (1935). If one takes the friction coefficient as a linear function of the spring elongation in the dumbbell model of Reference [7] (in the limt of $N \to \infty$), the Fokker-Planck equation for the distribution function of the polymer dumbbells is identical to that of liquid nuclei in the problem of vapor nucleation!

13. Y. Rabin, J. Polym. Sci.; Polym. Lett. **23**, 11 (1985).

14. R.B. Bird, R.C. Armstrong and O. Hassanger, Dynamics of Polymeric Liquids, Vol. 1 (Wiley, New York, 1977).

PROPAGATION OF DISCONTINUITIES IN LINEAR VISCOELASTICITY

Michael Renardy

Department of Mathematics
Virginia Polytechnic Institute and State University
Blacksburg, Virginia 24061

and

William J. Hrusa

Department of Mathematics
Carnegie-Mellon University
Pittsburgh, Pennsylvania 15213

Abstract:

We study the initial value problem in linear viscoelasticity with a step-jump in the initial data when the memory function has a singularity at 0. Such models show interesting intermediate behavior between hyperbolic equations, which propagate singularities, and parabolic equations, which smooth them out. In particular, we find coexistence of finite wave speeds and smoothing of the singularity. The degree of smoothing depends on the nature of the singularity in the viscoelastic memory function. A number of examples are discussed.

I. Introduction

This paper gives a summary of the results derived in [1] and [2]. We study the equation of motion of a one-dimensional viscoelastic medium occupying the whole real line

$$u_{tt}(x,t) = bu_{xx}(x,t) + \int_{-\infty}^{t} m(t-\tau)(u_{xx}(x,t) - u_{xx}(x,\tau))d\tau,\ x \in R,\ t > 0. \quad (1)$$

Here b is an nonnegative constant and m is a positive, monotone decreasing function (The material is a solid if $b > 0$ and a fluid if $b = 0$). For the equation to make sense, we have to assume that m is integrable at infinity and that $tm(t)$ is integrable near $t = 0$. We are interested in solutions developing from initial data with a discontinuity. More precisely, we study situations where u vanishes identically up to time 0,

$$u(x,t) = 0, \quad x \in R, \quad t < 0, \tag{2}$$

but the values of u and u_t as $t \to 0+$ are non-zero and have a jump

$$u(x,0) = \frac{1}{2}\alpha \operatorname{sgn} x, \; u_t(x,0) = \frac{1}{2}\beta \operatorname{sgn} x. \tag{3}$$

It is well known that the wave equation $u_{tt} = u_{xx}$ propagates singularities in the initial data with a finite speed. If m is smooth, then the equation of linear viscoelasticity shows the same behavior, but there is exponential damping of the amplitude of the discontinuities, like for the telegraph equation [3],[4],[5]. On the other hand, a parabolic equation like $u_t = u_{xx}$ has infinite speed of propagation and smooths discontinuities in the initial data. We note that if in the equation above we formally set m equal to minus the derivative of the δ-function, then the integral term becomes u_{xxt}, i.e. the equation (for $b = 0$) is just the time derivative of $u_t = u_{xx}$.

Here, we are interested in problems intermediate between these extremes. The kernel m has a singularity at 0, but not as strong as the derivative of the δ-function. These problems lead to interesting regularity properties of the solutions, intermediate between the hyperbolic and parabolic case. As we shall see, the speed of wave propagation is finite if m is integrable. However, if $m(t) \to \infty$ as $t \to 0$, then, instead of a jump discontinuity across the propagating wave front, there is smoothing in a degree depending on the nature of the singularity in m [1],[2],[6]. Examples for various choices of m are given below. Independent of whether m is singular or not, there is also a stationary singularity at the position of the original step jump in the initial data (except in a very special case).

These results raise interesting questions for the nonlinear case. It has been shown that models of nonlinear viscoelasticity with smooth kernels lead to development of singularities (shocks) from smooth, but large, initial data (see the papers by C. Dafermos and by J.A. Nohel and M. Renardy in these proceedings and the references cited in those papers). Since a singularity in the kernel suppresses the propagation of singularities, it may also be expected to suppress their development. At present no results are known on this problem except for small data.

2. Results for Regular Kernels

The problem (1)-(3) can be solved by using Laplace transform with respect to time. If we set

$$\hat{u}(x,\lambda) = \int_0^\infty e^{-\lambda t} u(x,t)dt, \tag{4}$$

we obtain the transformed equation

$$\lambda^2\hat{u}(x,\lambda) - \frac{1}{2}\alpha\lambda \operatorname{sgn} x - \frac{1}{2}\beta \operatorname{sgn} x = (b + \hat{m}(0) - \hat{m}(\lambda))\hat{u}_{xx}(x,\lambda). \tag{5}$$

For the case of singular kernels not integrable near 0, $\hat{m}(0) - \hat{m}(\lambda)$ should be interpreted to mean $\int_0^\infty m(t)(1 - e^{-\lambda t})dt$. Equation (5) for fixed λ is an ordinary differential equation in x, which has the solution

$$\hat{u}(x,\lambda) = \frac{\alpha\lambda + \beta}{2\lambda^2}\{\operatorname{sgn} x - H(x)\exp(-\lambda x/\sqrt{b + \hat{m}(0) - \hat{m}(\lambda)}\,)$$

$$+ H(-x)\exp(\lambda x/\sqrt{b + \hat{m}(0) - \hat{m}(\lambda)}\,)\}. \tag{6}$$

Here H denotes the Heaviside step function. By using the inversion formula for the Laplace transform, we obtain the following expression for $x,t > 0$ (for $x < 0$, we have $u(x,t) = -u(-x,t)$)

$$u(x,t) = \frac{1}{2}(\alpha + \beta t) - \frac{1}{4\pi i}\int_{\gamma-i\infty}^{\gamma+i\infty}\left(\frac{\alpha}{\lambda} + \frac{\beta}{\lambda^2}\right)\exp(\lambda t - \lambda x/\sqrt{b + \hat{m}(0) - \hat{m}(\lambda)}\,)d\lambda, \tag{7}$$

where γ is any positive number, and the square root is the branch that takes positive values on the real axis. It is clear from this expression that the regularity properties of u are determined by the behavior of $\hat{m}$ as $\lambda \to \infty$, which is more or less related to the behavior of m as $t \to 0$.

If m is integrable, then, according to the Riemann-Lebesgue lemma, $\hat{m}(\lambda)$ tends to 0 as $\lambda \to \infty$. As a consequence, the integral in (8) can be evaluated be deforming the contour into the right half plane if $x > t\sqrt{b + \hat{m}(0)}$, and turns out to be zero. That is, the solution simply extrapolates the initial data for $x > 0$ and has not yet felt the jump in the initial data at $x = 0$. We have a finite speed of wave propagation.

If we assume that m is very smooth, i.e. that $m \varepsilon C^{\infty}[0,\infty)$ and that all deriviatves of m are integrable, then $\hat{m}(\lambda)$ can be expanded in an asymptotic power seriesin $\frac{1}{\lambda}$. This leads to a corresponding asymptotic expansion for the integrand in (7). Each term has its only singularity at $x = t\sqrt{b + \hat{m}(0)}$, and the terms in the integrand decay with higher and higher powers of λ, leading to smoother and smoother contributions to u. It can be concluded from this that u is of class C^{∞} away from the line $x = t\sqrt{b + \hat{m}(0)}$. Across the line $x = t\sqrt{b + \hat{m}(0)}$, u (or, if $\alpha = 0$, its first derivatives) has a jump of exponentially decreasing amplitude (see [3-5]).

The expression (6) is also singular across the line $x = 0$. In fact, if we set

$$J(t) = u_{xx}(0^+,t) - u_{xx}(0^-,t), \tag{8}$$

we find from (6)

$$\hat{J}(\lambda) = -\frac{\alpha\lambda + \beta}{b + \hat{m}(0) - \hat{m}(\lambda)}. \tag{9}$$

Lower derivatives of u can be shown to be continuous across $x = 0$. An analogue of this stationary singularity appears in the study of the Riemann problem [7-9]. The jump $J(t)$ vanishes for positive t if and only if $\hat{J}(\lambda)$ is a polynomial. It is easy to show that this is possible only if m is an exponential. In [2], we also show some results on the behavior of J as $t \to \infty$. If $b > 0$ or $b = \beta = 0$ and $tm(t)$ is integrable at infinity, then $J(t) \to 0$ as $t \to \infty$. If, on the other hand, $b = 0$, $\beta \neq 0$ and we assume $t^2m(t)$ is integrable at infinity, then J tends to a nonzero constant. These results are proved using a theorem of Jordan, Staffans and Wheeler [10].

3. Examples of Singular Kernels

We now turn to the study of kernels which have the property $m(t) \to \infty$ as $t \to 0$. If m is integrable near 0, then we still have a finite wave speed. However, the formula for the amplitude of the jump across the wave in the case of regular kernels [3-5] involves $m(0)$, and it gives zero if $M(0) = \infty$. This indicates that there is some smoothing of the solution at $x = \pm t\sqrt{b + \hat{m}(0)}$. Examples are studied below. As in the case of regular kernels, there is a jump in u_{xx} at the line $x = 0$. If $m(t)$ is C^∞ away from $t = 0$, it is clearly to be expected that the solution will be C^∞ away from $x = 0$ and $x = \pm t\sqrt{b + \hat{m}(0)}$, but we have not been able to prove this in full generality. A weaker result, assuming growth conditions on m and its derivatives near $t = 0$, is proved in [2]. Moreover, if m is completely monotone, as it is in all the examples discussed below, then it is not difficult to show that the solution is analytic away from $x = 0$ and $x = \pm t\sqrt{b + \hat{m}(0)}$.

The regularity of u across $x = t\sqrt{b + \hat{m}(0)}$ is determined by the decay of the integrand in (7) as $\lambda \to \infty$, and hence by the asymptotic behavior of $\hat{m}(\lambda)$. Although there is no sharp equivalence, the behavior of $\hat{m}$ at infinity is more or less related to the behavior of m at 0. As the following examples show, the smoothing of the solution becomes stronger as the strength of the singularity in m increases.

The kernel

$$m(t) = \sum_{n=1}^{\infty} e^{-n^p t}, \quad p > \frac{1}{2} \tag{10}$$

behaves like $t^{-1/p}$ as $t \to 0$, but decays exponentially at infinity. The asymptotic behavior of its transform is given by (see [1])

$$\hat{m}(\lambda) \sim \lambda^{1/p-1} \frac{\pi}{p} \operatorname{cosec} \frac{\pi}{p}, \tag{11}$$

if $p > 1$. If $p < 1$, then the kernel is not integrable, and we have

$$\hat{m}(0) - \hat{m}(\lambda) \sim \ln(1 + \lambda) + C, \quad p = 1 \tag{12}$$

$$\hat{m}(0) - \hat{m}(\lambda) \sim \lambda^{1/p-1} \frac{\pi}{p} \operatorname{cosec} \frac{(1-p)\pi}{p}, \quad p < 1. \tag{13}$$

In all these cases, the integrand in (7) decays faster than any power of λ as $\lambda \to \infty$. As a consequence, we can differentiate the integral with respect to x or t an arbitrary number of times, and we still obtain a convergent integral. Hence u is of class C^∞ in the quadrant $x > 0$, $t > 0$. If $p < 1$, the wave speed is infinite and u is analytic for $x > 0$, $t > 0$, but for $p > 1$ the wave speed is finite. Hence smoothing coexists with a finite wave speed.

The kernel

$$m(t) = \sum_{n=0}^{\infty} e^{-e^n t} \tag{14}$$

has a logarithmic singularity at $t = 0$, and its transform can be shown to behave like

$$\hat{m}(\lambda) \sim \frac{\ln(1+\lambda)}{\lambda} \tag{15}$$

as $\lambda \to \infty$. The integrand in (7) decays with a power of λ proportional to x and hence the regularity of the solution across the wave increases proportional with x (or equivalently, proportional with t).

A still weaker singularity is given by the kernel

$$m(t) = \sum_{n=1}^{\infty} e^{-te^{e^n}}. \tag{16}$$

As $t \to 0$, m behaves like $\ln|\ln t|$, and the leading term in the transform is

$$\hat{m}(\lambda) \sim \frac{\ln \ln \lambda}{\lambda}. \tag{17}$$

The integrand in (7) decays like $\frac{1}{\lambda}$ times an inverse power of $\ln \lambda$, which increases proportional to x. This implies continuity across the wave for large enough x, but the gain in regularity (expressed in terms of the number of existing derivatives) is only "infinitesimal". We have a conjecture that if m is singular and maybe satisfies some monotonicity conditions, then u should be continuous for any $t > 0$, but we have not been able to prove this even for the

m above. Using more detailed asymptotic results on $\hat{m}$, we were able to establish such a result when the sum in m is replaced by an integral, i.e. for the kernel

$$m(t) = \int_0^\infty \exp(-te^{e^{\nu}})d\nu. \tag{18}$$

Acknowledgement

This research was sponsored by the United States Army under Contract No. DAAG-29-80-C-0041 and supported in part by the National Science Foundation under Grants No. MCS-8215064 and MCS-8210950.

References

[1] M. Renardy, Some remarks on the propagation and non-propagation of discontinuities in linearly viscoelastic liquids, Rheol. Acta 21 (1982), 251-254.

[2] M. Renardy and W.J. Hrusa, On wave propagation in linear viscoelasticity, Quart. Appl. Math., 43(1985), 237-254.

[3] B.T. Chu, Stress waves in isotropic linear viscoelastic materials, J. Mécanique I (1962), 439-446.

[4] G.M.C. Fisher and M.E. Gurtin, Wave propagation in the linear theory of viscoelasticity, Q. Appl. Math. 23 (1965), 257-263.

[5] B.D. Coleman, M.E. Gurtin and I.R. Herrera, Waves in materials with memory, Arch. Rat. Mech. Anal. 19 (1965), 1-19 and 239-265.

[6] K.B. Hannsgen and R.L. Wheeler, Behavior of the solutions of a Volterra equation as a parameter tends to infinity, J. Integral Eq. 7(1984), 229-237.

[7] J.M. Greenberg, L. Hsiao and R.C. MacCamy, A model Riemann problem for Volterra equations, in: Volterra and Functional Differential Equations, K.B. Hannsgen et al. (ed.), Dekker 1982, 25-43.

[8] J.M. Greenberg and L. Hsiao, The Riemann problem for the system $u_t + \sigma_x = 0$ and $(\sigma - f(u))_t + (\sigma - \mu f(u)) = 0$, Arch. Rat. Mech. Anal. 82 (1983), 87-108.

[9] R.C. MacCamy, A model Riemann problem for Volterra equations, Arch. Rat. Mech. Anal. 82 (1983), 71-86.

[10] G.S. Jordan, O.J. Staffans and R.L. Wheeler, Local analyticity in weighted L^1-spaces and applications to stability problems for Volterra equations, Trans. Amer. Math. Soc. 174 (1982), 749-782.

STRENGTH AND ENTANGLEMENT DEVELOPMENT AT AMORPHOUS POLYMER INTERFACES

R. P. Wool

Department of Materials Science
University of Illinois at Urbana-Champaign
1304 W. Green Street
Urbana, Illinois 61801

Abstract

When two similar amorphous dense polymers are brought into good contact at sufficiently high temperatures, we consider strength development resulting from interdiffusion of chain segments across the interface. Mechanical properties, H, are evaluated in terms of microscopic mechanisms involving chain segment slippage and bond rupture. Using deGennes' topological constraint reptation model for the interdiffusion analysis of the symmetric interface, we obtain the mechanical properties as a function of time, $t < t_\infty$, and molecular weight, M, as

$$H = H_\infty \left(\frac{t}{t_\infty} \right)^{r/4} \qquad r,s = 0, \pm 1, \pm 2, \ldots$$

where

$$H_\infty = M^{(3r-s)/4}$$

$t_\infty \sim M^3$ is a characteristic relaxation time, H_∞ is the virgin state property, and r, s have integer values. For fracture stress, $r = 1$, for fracture energy, G_{IC}, $r = 2$, for fatique crack propagation rate da/dn, $r = -5$, for tensile modulus, $r = 0$. The value of H_∞ depends on the microscopic deformation process such that when bond rupture dominates, s is determined by $s = 3r$ and when chain slippage dominates, $s = r$. Experimental evidence for these relations is presented.

Paper presented at Workshop on Amorphous Polymers and Non-Newtonian Fluids, March 4-8, 1985, Institute for Mathematics and Its Applications, University of Minnesota

The role of entanglements in amorphous polymer mechanical properties is explored. The onset of entanglement at a critical molecular weight, M_c, is examined in terms of a molecular bridge model. The model suggests that when each chain is long enough on average to form a single S-shaped bridge, a connected structure develops and M_c is determined by

$$M_c = \frac{30.89\, C_\infty M_0}{\alpha^2 j c}$$

where C_∞, M_0, j, α and c are the characteristic ratio for a Gaussian chain, the monomer molecular weight, the number of bonds per monomer, the ratio of the chain contour relaxed length to the length along the bonds (~ 0.83) and the solution concentration, respectively. Reasonable agreement with experiemental values is obtained.

Finally, the reptation model used in the interface study is examined by new experimental methods. Upon subjecting an amorphous polymer to a step function strain, the reptation model predicts that the relaxation by diffusion of oriented segments or dipoles should behave approximately as

$$F = 1 - at^{1/2} M^{-3/2}$$

where F is the normalized Herman's orientation function of the polymer segments and a is a constant. F was measured experimentally by infrared dichoroism methods in monodisperse molecular weight samples $(M >> M_c)$ of polystyrene. These results and studies on centrally deuterated polystyrene chains provide support for the reptation model.

Introduction

The Problem

When two similar amorphous polymers make good contact to form a polymer-polymer interface, we ask how strength develops as a function of contact time, t, and molecular weight, M , of the polymers. Fracture stress, σ , shear stress, τ , critical fracture energy, G_{IC} , and fatique crack propagation rates, da/dn , are determined as functions of t and M . Solutions to this problem have application to polymer processing, internal weld-lines, lamination and the physics of fracture mechanics. Contributions to solutions have been recently given by Prager and Tirrell [1,2], deGennes [3,4], Kausch [5,] and by Wool et al. [7,8,9].

Solutions

In our approach we divide the problem into three parts. The first part consists of determining a set of molecular properties for the interface, H(t). The second part relates the properties of the interface to the mechanical properties via a set of deformation mechanisms involving chain disentanglement and bond rupture. The third part of the solution consists of determining the fracture mechanics of a crack propagating through an interface using solutions of the first two parts. This method of solution involves the evaluation of three different but interconnected problems and results in an understanding of the fracture mechanics of polymers, in terms of the physics and chemistry of the interface.

Polymer-Polymer Inderdiffusion

The properties of the interface were determined using the molecular dynamics model of deGennes [10] and Doi-Edwards [11]. Our version of the reptation model is shown in Fig. 1, as developed by Kim and Wool [9] for polymer-polymer interdiffusion. Following contact at $t=0$, the chains interdiffuse across the interface as shown in Fig. 2. For this case, we determine a set of properties, H as a function of time, t , and molecular weight, M . They include the number of chains, n , number of bridges, p , average monomer interpenetration depth, X , the total monomer depth, X_0 , the center of mass diffusion, X_{cm} , the average contour interpenetration length, l, the total contour length, L_0 , and the average bridge length, l_p . The time dependent, H(t), and equilibrium solutions, H_∞ , can be summarized as [12]

$$H(t) \sim t^{r/4}M^{-s/4};\ H_\infty\ M^{3r-s)/4} \quad \text{, where } r, s = 1,2,3,4 \ldots \qquad (1)$$

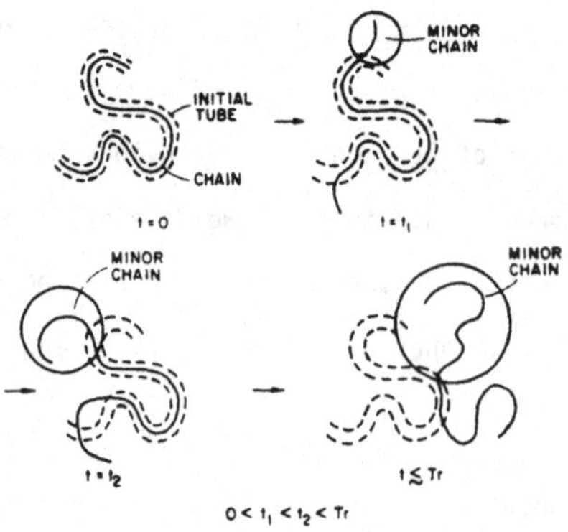

Fig. 1. The reptation model for a chain in an entangled melt is shown relaxing from its tube at several times $t < T_r$, where T_r is the reptation time. The part of the chain which has relaxed from its original tube is termed the minor chain and is shown in its most probable random-coil spherical envelope.

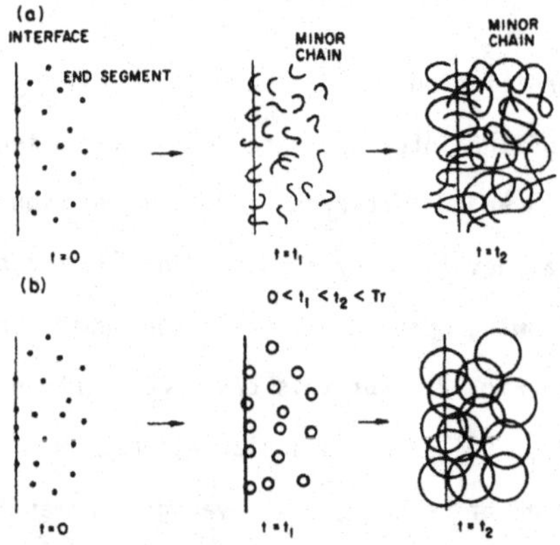

Fig. 2. (a) Interdiffusion of the minor chains at the polymer-polymer interface (one side of the interface is shown for convenience) is shown at times $t < T_r$. (b) The behavior of the minor chains is shown during the interdiffusion process.

or as a scaling law,

$$H(t) \approx H_{\infty}(t/t_{\infty})^{r/4} \qquad (2)$$

where $t_{\infty} \sim M^3$. the set of properties in terms of integers $H(r,s)$ are as follows: $n(1,5)$, $p(2,6)$, $X(1,1)$, $X_0(2,4)$, $L(2,2)$, $L_0(3,7)$ and $1_p(1.1)$.

Table I summarizes the molecular aspects of interdiffusion at a polymer-polymer interface. These relations can be used to test hypotheses of strength development. If a mechanical property is directly related to $H(t)$ or a product of several $H(t)$ properties, then the time dependence, molecular weight dependence of healing and the molecular weight dependence of the virgin (healed) state must be simultaneously predicted. This approach suggests critical experiments to be performed to test models for healing [1-9] and provides a method of testing virgin state theories of strength.

Table 1. Molecular aspects of interdiffusion at a polymer-polymer interface.

Molecular Aspect	Symbol	Dynamic Relation, H(t)	Static Relation
Number of Chains	$n(t)$	$t^{1/4}M^{-5/4}$	$M^{-1/2}$
Number of Bridges	$p(t)$	$t^{1/2}M^{-3/2}$	M^0
No. of Monomers	$L_0(t)$	$t^{3/4}M^{-7/4}$	$M^{1/2}$
Average Depth	$X(t)$	$t^{1/4}M^{-1/4}$	$M^{1/2}$
Total Depth	$X_0(t)$	$t^{1/2}M^{-3/2}$	M^0
Center of Mass	$X_{cm}(t)$	$t^{1/2}M^{-1}$	$M^{1/2}$
Average Length	$\ell(t)$	$t^{1/2}M^{-1/2}$	M
Average Bridge Length	$\ell_p(t)$	$t^{1/4}M^{-1/4}$	$M^{1/2}$
General Property	$H(t)$	$t^{r/4}M^{-s/4}$ r,s = 1,2,3,4...	$M^{(3r-s)/4}$

Microscopic Deformation Mechanisms

When chain disentanglement dominates the strength of the interface, a microscopic fracture analysis involving the strain energy dissipated in chain pull-out [13] results in $\sigma \sim X(t)$. From Table 1, substituting for $X(t)$, we have the following results:

$$\sigma(t) \sim t^{1/4} \qquad (t < t_\infty) \; .$$

$$\sigma \sim M^{-1/4} \qquad (\text{constant } t) \qquad (3)$$

$$\sigma_\infty \sim M^{1/2} \qquad (t > t_\infty)$$

$$\sigma \sim \gamma^{1/2} \qquad (\text{rate effect})$$

These results are demonstrated for several polymers in Figs. 3, 4 and 5.

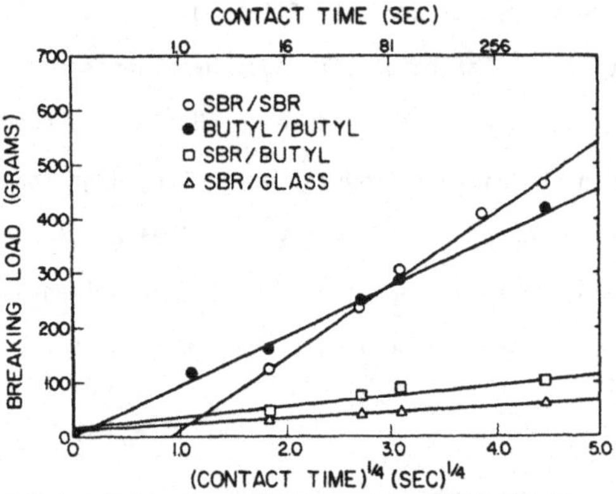

Fig. 3. The uniaxial breaking load (Skewis data [13(b)]) for several linear uncured polymer-polymer pairs is shown versus contact time, $t^{1/4}$. Details in ref. 13. The styrene-butadiene (SBR)/Butyl pair form an asymmetric incompatible interface resulting in low strength developed at long times.

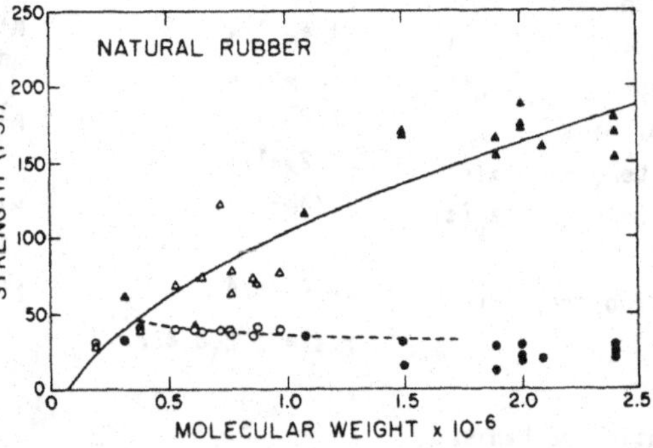

Fig. 4. Tack (circles) and green strength (triangles) are shown as a function of molecular weight for fractionated samples of natural rubber . Data obtained by Forbes and McLeod [13(c)]. The dashed line for strength (tack) at constant time represents the relation, $\sigma \sim M^{-1/4}$, for healing and the solid line for the virgin (green) strength represents $\sigma_\infty \sim M^{1/2}$, as predicted by the chain disentanglement model.

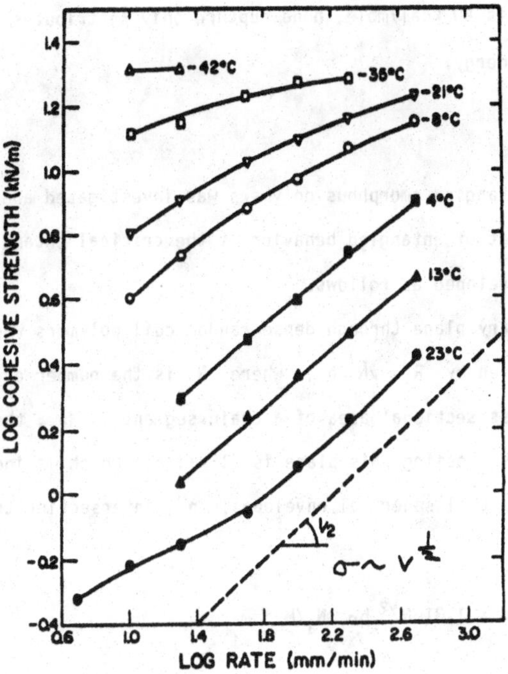

Fig. 5. The cohesive strength of an SBR rubber is shown versus peeling rate in a T-peel test. Data of Hamed [13(d)]. The slope of 1/2 is shown as predicted by the disentanglement model [13(a)].

The fracture of glassy polymers (healed above the glass transition temperature) is more complex due to the role of bond rupture and the mechanism of a crack propagating through a craze-like zone.

Molecular fracture in amorphous polymers produced by a crack propagating through a craze microstructure was simulated by a microtome slicing method [14]. A GPC analysis of the polystyrene slices gave exact measures of N_f, the number of broken bonds per m^2, N_f, was found to be [15]

$$N_f = \frac{7 \times 10^{17}}{m^2} \quad M^0 \tag{4}$$

At room temperature, N_f was independent of M but at higher fracture temperatures, N_+ decreases and is molecular weight dependent. Assuming that the energy to break a bond is 80 kcal/mole, bond rupture only contributes a factor of 10^{-4} to the fracture energy.

Polymer Entanglements

The topology of entangled amorphous polymers was investigated and a description for the onset of entangled behavior at the critical entanglement molecular weight was developed as follows.

Consider an arbitrary plane through dense random coil polymers whose r.m.s. end-to-end vector is given by $R = \sqrt{N}\ b$, where N is the number of steps of length b. If the cross sectional area of a chain segment is a, the total number of chain segments crossing this plane is $1 = 1/a$, which is independent of N. The number of random coil spherical envelopes, n, intersecting this plane is given by [14]

$$n = 1.31\ N^{1/2}\ b\rho\ N_a/M \tag{5}$$

where ρ is the density, N_a is Avogadros number and M is the molecular weight. Since $N \sim M$, we have $n \sim M^{-1/2}$. We can examine structure development at the plane by defining the average number of intersects per chain, p_c, as

$$p_c = I/n = 1/an \tag{6}$$

At very low M, $p_c = 1$ as shown in Fig. 6 and the system behaves as a fluid without entanglements. As M increases each chain eventually intersects the plane twice, such that $p_c = 2$ and the average structure formed is that of a loop which may be capable of forming at least a partial entanglement. As M increases further, we obtain $p_c = 3$ which produces a bridge-like structure. Since every plane in the polymer possesses this average structure we now have a fully interconnected state. When $p_c < 3$, the structures can slip apart rather easily but when $p_c > 3$ the interconnected structures can only separate from each other by self diffusion or fracture mechanisms until they become disconnected. This approach is similar to a percolation arguement at the critical threshold.

A

B

C

D

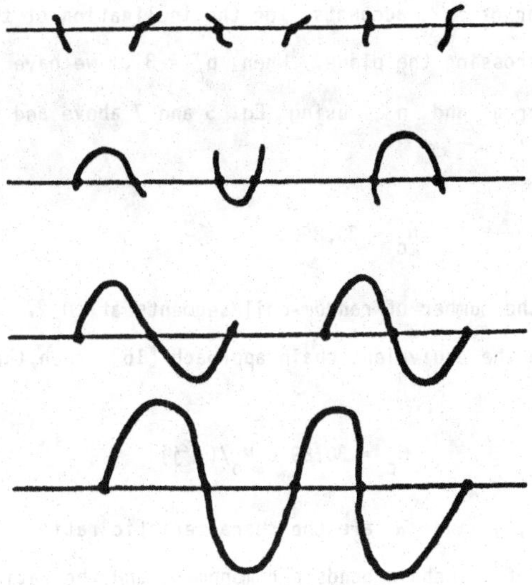

Fig. 6. The development of bridge structure with increasing molecular weight is shown for random coil chains intersecting an arbitrary plane. (a) At low M_1 each chain intersects the plane one time on average. (b) At $M_2 > M_1$, the average structure is that of a loop where each chain intersects twice. (c) At $M_3 > M_2$, each chain intersects the plane three times and bridges are formed. The transition from loops to bridges occurs at $M_3 = 2.25\ M_2$. The bridge structure of M_3 is considered the elementary structure in forming a fully connected entangled network. (d) At M_6, a highly entangled chain is shown. Note that the number of crossings is independent of M but the number of chains intersecting the plane behaves as $n \sim M^{-1/2}$ and the average bridge length increases as $M^{1/2}$.

Critical Entanglement Molecular Weight, M_c

We now explore the onset of a connected state at $p_c = 3$ in terms of M_c, where we assume for now that $M_3 = M_c$. The cross sectional area a, can be readily obtained by dividing the volume of a chain by its length, Nb, as

$$a = \frac{\sqrt{2}\ M/(\rho N_a)}{Nb} \tag{7}$$

where the factor of $\sqrt{2}$ accounts for the inclination of the randomly oriented chain segment crossing the plane. When $p_c = 3$, we have $3an = 1$ and substituting for a and n, using Eq. 5 and 7 above and solving for $N = N_c$, we obtain

$$N_c = 30.89 \tag{8}$$

where N_c is the number of random-coil segments at M_c. To convert from N_c to M_c, we use the equivalent chain approach [16] such that M_c is given by,

$$M_c = 30.89\, C_\infty M_0/(\alpha^2 j) \tag{9}$$

where C_∞, M_0, j and α are the characteristic ratio, monomer molecular weight, number of mainchain bonds per monomer, and the ratio of the relaxed chain length to the bond contour length, respectively. In this case α depends on the chain conformation and can be determined for crystalline polymers from unit cell dimensions as,

$$\alpha = C/(b_0 z j) \tag{10}$$

where C is the c-axis unit cell dimension, z is the number of monomers per chain per C-axis length and b_0 is the real bond length ($b_0 \approx 1.54$ Å). Thus M_c can also be determined from the following,

$$M_c = 30.89\, (b_0\, z/C)^2\, C_\infty M_0 j \tag{11}$$

Equations (9) and (11) describe the onset of bridge structures at molecular weight, M_3. If $M_3 = M_c$, then Eqns. (9) and (11) should predict M_c values for all linear amorphous polymers. It is interesting to note that the theoretical description of M_c contains no adjustable parameters and the parameters for each polymer are either available in the literature or can be readily estimated. Using Eq. (11), we calculated M_c and compared results with several polymers of known M_c values obtained from references 17 and 18. The parameters C_∞ and the C-axis dimension were obtained from references 17 and 19. Table 2 shows a comparison of theoretical and experiemental M_c values for several linear polymers and the agreement is considered to be excellent.

This model for M_c which employs connectivity between static bridge-like structures, resulting in $N_c = 30.89$ compares with J. Klein's [20] value of $N_c = 13.33$ which was deduced from dynamic considerations of the Rouse to reptation transition in the zero-shear viscosity. His value of N_c corresponds to the point where chain diffusion becomes restricted to reptation. Kremer [21] using a numerical analysis of the statics and dynamics of polymer melts determined the number of random coil segments between entanglements, N_e, as 13 ± 3, 50 ± 10, 20 ± 5 and 9 ± 1, depending on the time region examined. If N_c is roughly twice N_e, then this range of values is consistent with our value of $N_c \approx 30$ and $N_c/2 \approx 15$. Further support for the value of $N_c = 30$ can be obtained from bridge statistics and computer studies of random walks [22].

In Eq. (5), $\alpha^2 j \approx 1$ for many vinyl polymers and a useful approximation for M_c is obtained as,

$$M_c \approx 30\, M_o C_\infty \tag{12}$$

For example, using values for M_o and C_∞ from Table 2 we see that Eq. (12) gives M_c (polystyrene) $\approx 31{,}200$ and M_c (Polypropylene) $\approx 7{,}800$, in good agreement with experiment. Equation (12) emphasizes the proportionality, $M_c \sim C_\infty$, which is in agreement with Kleins theory [20] but disagrees with Aharoni [23] who found that $M_c \sim C_\infty^2$ from numerical analysis of published M_c and M_e values, although considerable scatter occurred in his data plots.

Table II

Polymer	M_o	C_∞	j	b_o °A	C °A	z	M_c (theory)	Mc (expt)
Polyethylene	28	6.7	2	1.54	2.55	1	4,000	3,800
Polystyrene	104	10	2	1.54	6.5	3	32,000	31,200
Polypropylene	42	6.2	2	1.54	6.5	3	8,000	7,000
Polyvinylalcohol	44	8.3	2	1.54	5.51	2	7,000	7,000
Polyvinylacetate	86	9.4	2	1.54	6.5[a]	3	25,000	24,500

[a]Assumed 3_1 helix

Disentanglement and Fracture

In uniaxial deformation experiments, disentanglement can occur with increasing draw ratio, λ . When $M > M_c$, the bridge model predicts that the critical draw ratio for disentanglement, λ_c , occurs at

$$\lambda_c = \sqrt{\frac{M}{M_c}} \tag{13}$$

At λ_c , the connectivity between bridges becomes that of $p_c = 3$ as shown in Fig. 6. This result is important in crazing and fracture studies where disentanglement via chain pullout is an important mechanism in addition to bond rupture. We have recently shown [15] that the disentanglement mechanism contributes about 99.99% of the fracture energy, G_{IC}, in polystyrene. Using GPC methods on slice-fractured monodisperse polystyrene samples, the number of broken bonds obtained by a crack propagating through a craze was measured at $6 \times 10^{13}/cm^2$ which corresponds to a fracture energy of about 10^{-4} kJ/m^2. However, the experimental fracture energy is of the order of 1 kJ/m^2 . When bond rupture dominates the fracture process, then $\lambda_c \sim M^0$ and $G_{IC} \sim M^0$. The role of bond rupture is to largely control the extent of disentanglement and energy dissipation.

The critical strain at fracture, ε_c is determined via $\varepsilon_c = \lambda_c - 1$ as

$$\varepsilon_c \sim M^{1/2} - M_c^{1/2} \tag{14}$$

which is in excellent agreement with observations by Wool and O'Connor on fracture of low molecular weight $(M > M_c)$ linear amorphous polymers as shown in Fig. 8. When the crack opening displacement, δ , is proportional to ε_c , then it follows that $\delta \sim M^{1/2} - M_c^{1/2}$. The latter result is consistent with results of Pitman and Ward [26] for compact tension fracture mechanics studies of low molecular weight polycarbonates $(M > M_c)$. When the critical stress, σ_c , is controlled by chain pullout, then $\sigma_c \sim M^{1/2} - M_c^{1/2}$ such that the critical stress intensity factor, K_{IC} , also behaves as $K_{IC} \sim M^{1/2} - M_c^{1/2}$, and the critical strain energy release rate behaves as $G_{IC} = \sigma_c \delta \sim (M^{1/2} - M_c^{1/2})^2$ When $M >> M_c$, then $G_{IC} \sim M$ if disentanglement dominates the fracture process.

Fracture Mechanisms of Interfaces

Wedge-Cleavage methods were used to study the time and molecular weight dependence on welding. For mixed mechanism (chain pullout and bond rupture) we find that the Dugdale fracture mechanics model predicts [27] for glassy polymers,

$$
\begin{aligned}
&G_{IC} \sim t^{1/2} M^{-x} && 1/2 < x < 1 \\
&G_{IC} \sim (M - M_c) && M_c < M < M^* \\
&G_{IC} \sim M^0 && M > M^* \\
&\left(\frac{\partial a}{\partial N}\right)_{\Delta K} \sim M^{-5/2} &&
\end{aligned}
\tag{15}
$$

where $M^* \approx 2 \times 10^5$ for polystyrene at room temperature and increases with T. Since $K_{IC} \sim G_{IC}$, the dependence of the critical stress intensity factor on t and M can be readily derived. Experimental support for these relations is given in Figs. 7-9 [27,28,29]. The value of X does not agree with that predicted in Refs. [1-6].

Infrared Studies of Reptation Dynamics

Further support for the reptation model as used in Fig 1 has recently been obtained by us [30] using Fourier Transform Infrared Spectroscopy, FTIR. The Hermans orientation function, F, was measured as a function of t and M for uniaxially step-strained monodisperse molecular weight polystyrene melts. In Fig. 1, note that the minor chain which has excaped from its tube loses its orientation such that the chain orientation function is simply the fraction of chain remaining in the tube and is given by

$$F = 1 - \frac{4}{\pi^{3/2}} \left(\frac{t}{T_r}\right)^{1/2} \qquad (t < T_r)$$

where T_r is the reptation time. Since $T_r \sim M^3$, the time and molecular weight dependence of the orientation function is obtained during stress relaxation as,

$$F = 1 - \alpha t^{1/2} M^{-3/2} \qquad (t < T_r)$$

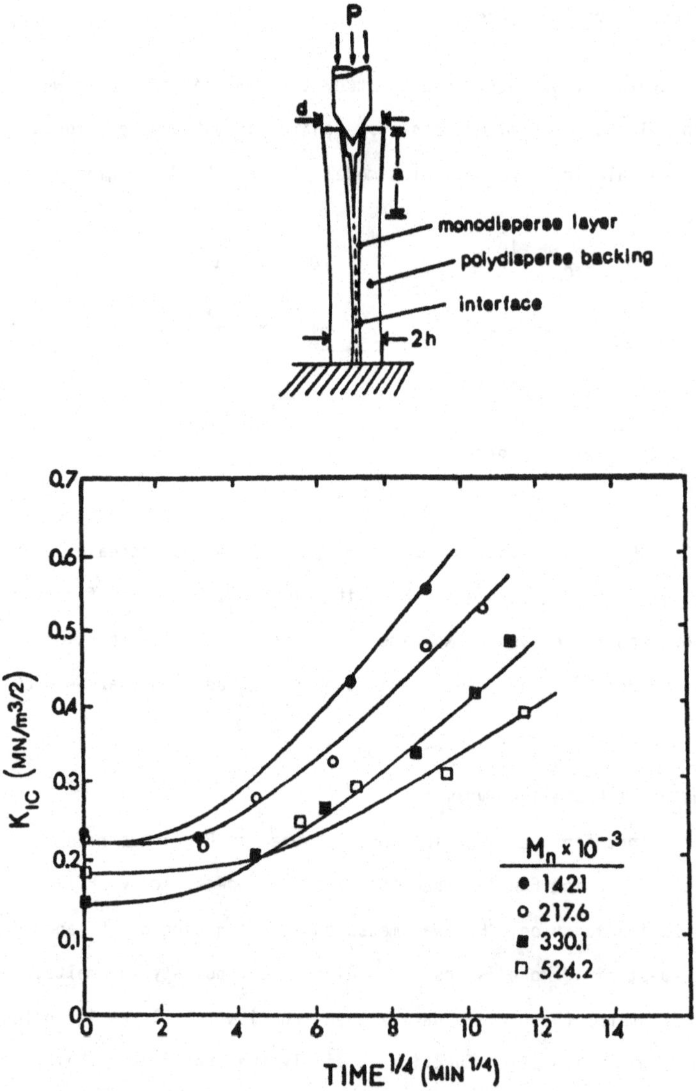

Fig. 7. The Wedge cleavage method for determining the strength of a healing interface is shown (top). The critical stress intensity factor, K_{IC}, is plotted vs time $t^{1/4}$ for polystyrene interfaces of indicated monodisperse molecular weights. The interface was wetted at 135°C for 30 sec and then healed at 120°C for the times shown. At long times the data is well described by $K_{IC} \sim t^{1/4} M^{-x}$ where $x \approx 1/2$.

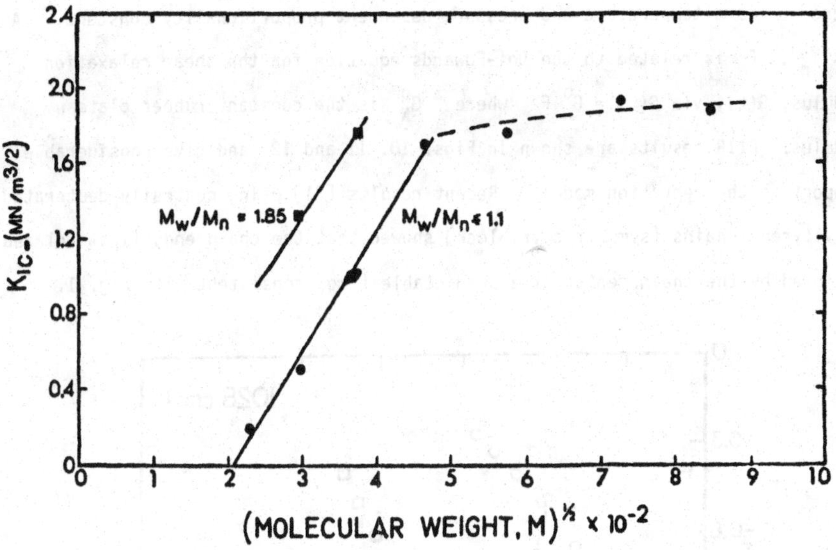

Fig. 8. the critical stress intensity factor, K_{IC}, is plotted versus molecular weight $M^{1/2}$ for virgin compact tension samples of polystyrene, (squares polydisperse, circles, monodisperse molecular weight). the results show that $K_{IC} \sim M^{1/2} - M_c^{1/2}$ in the range $M_c < M < 2 \times 10^5$ [29].

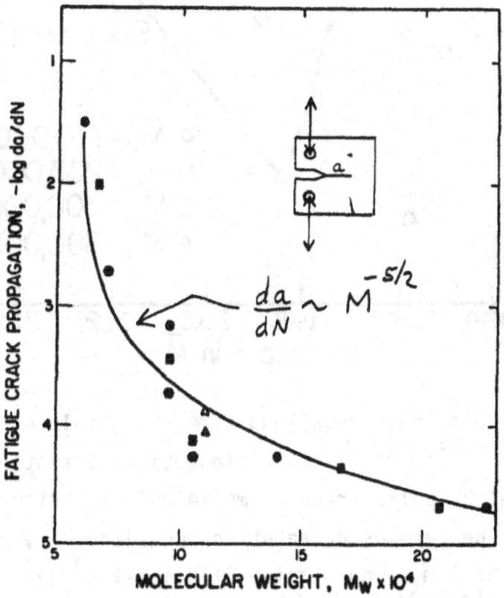

Fig. 9. Fatique crack propagation rates vs molecular weight are shown for PVC (data of Rimnac et al. [28]). The solid line is given by chain pullout theory as $da/dN = \alpha/(M^{5/2} - M_{Cr}^{5/2})$ where $\alpha = 4.5 \times 10^8$ mm $M^{5/2}$ / cycle and $M_{Cr} = 6 \times 10^4$ is the molecular weight at which critical crack propagation occurs at $\Delta K = 0.7$ MPa $\sqrt{m}$.

where α is a temperature and pressure dependent proportionality constant. At $t < T_r$, F is related to the Doi-Edwards equation for the shear relaxation modulus, G(t), via $G(t) = G_N^0 F$ where G_N^0 is the constant rubber plateau modulus. FTIR results are shown in Figs. 10, 11 and 12 and give considerable support to the reptation model. Recent results [31] using centrally deuterated polystyrene chains [symmetric triblock] showed that the chain ends first relaxed followed by the chain center at a predictable time, consistent with Fig. 1.

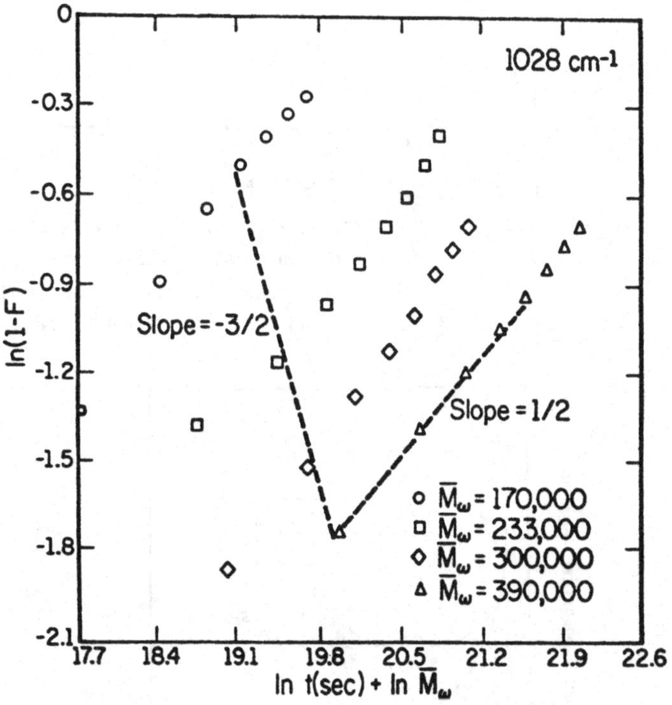

Fig. 10 Log (1-F) is plotted versus long time + log molecular weight to test the equation $1\text{-}F = t^{1/2}\ \bar{M}^{-3/2}$ for orientation relaxation of monodisperse molecular weight polystyrene films in uniaxial step strain of 300% at t = 115°C. The Hermans orientation function, F, was determined by infrared dichroism using Nicolet FTIR model 170-SX. The results show that $1\text{-}F \sim t^{1/2} M^{-3/2}$ at times short compared to terminal relaxation times. The data was obtained from the 1028 cm^{-1} band associated with the in-plane CH bending mode of the aromatic ring. Nearly identical results were obtained for the 906 cm^{-1} and 2850 cm^{-1} bands [30].

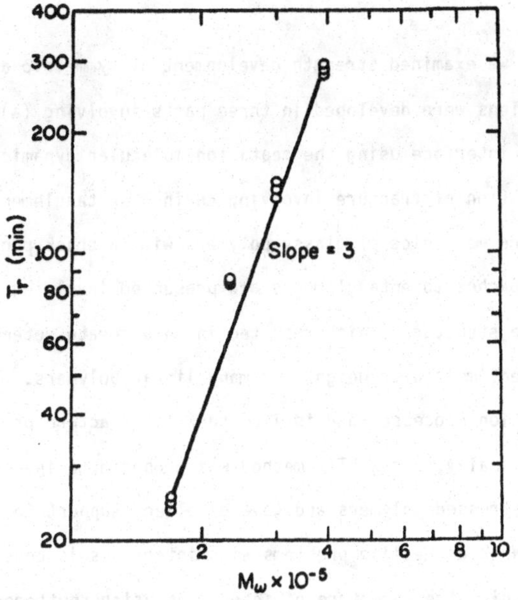

Fig. 11. The log of the relaxation times, t_r, obtained by infrared dichroism relaxation are plotted vs the log of molecular weight for monodisperse samples of polystyrene. The line drawn has a slope of 3.0. The three circles at each molecular weight represent data from three infrared bands in the same sample.

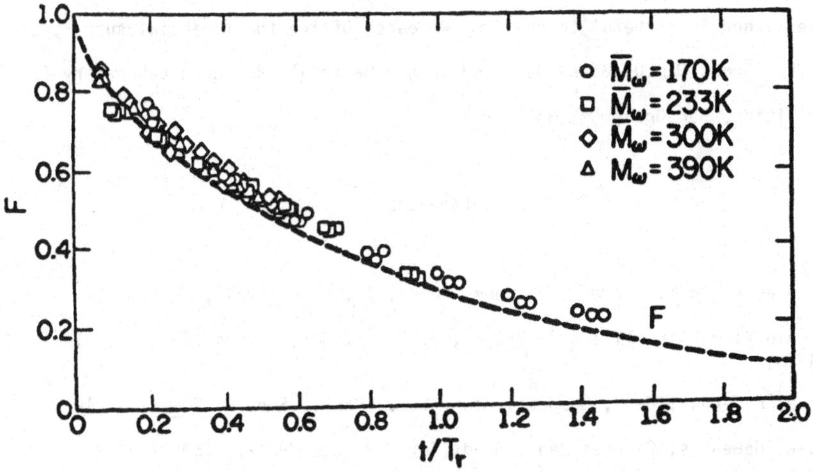

Fig. 12. The Hermans orientation function F is plotted versus t/T_r for the polystyrene samples of indicated monodisperse molecular weights. The dashed line represents the prediction of the Doi-Edwards theory [11],

$$F = \frac{8}{\pi^2} \sum_{j=0}^{\infty} \frac{1}{(2j+1)^2} \exp(-(2j+1)^2 \frac{t}{T_r})$$

Summary

In this paper we examined strength development at symmetric amorphous polymer interfaces. Solutions were developed in three parts involving (a) a molecular description of the interface using the reptation molecular dynamics model, (b) a microscopic description of fracture involving chain disentanglement and rupture, and (c) the fracture mechanics of glassy polymers with a crack propagating through a craze. New approaches to entanglements are presented involving connectivity between bridge-like structures which resulted in an accurate determination of the critical entanglement molecular weight for many linear polymers. This model is similar to percolation processes and is important for fracture processes involving disentanglement. Finally, a new FTIR method was proposed to investigate the molecular dynamics of stressed polymers and gave excellent support to the minor chain reptation model. Work on healing problems with interfaces is continuing with emphasis on determining the structure of interfaces using sputtered neutral atom spectrometry and evaluating the fractal nature of polymer interfaces.

Acknowledgement

The author is grateful to the Army Research Office for financial support of this work, Grant DAAL03-86-K-0034, and to the Materials Research Laboratory for its facilities, NSF Grant DMR-83-16 981.

References

1. S. Prager and M. Tirrell, J. Chem. Phys., 75, 5194 (1981).
2. D. Adolf, M. Tirrell and S. Prager, J. Polym. Sci. , Phys. Ed., 23, 413 (1985)
3. P.-G. deGennes, Hebd, Scances, Acad. Sci., Paris, Ser. B, 291, 219 (1980)
4. P.-G. deGennes, Comptes Rend. Acad. Sci., Paris, 292(2), 1505 (1981)
5. K. Jud, H. H. Kausch and J.A. Williams, J. Mat. Sci., 16, 204 (1981).
6. H.H. Kausch, I.U.P.A.C. Macromolecules, Ed. H. Benoit and P. Rempp, Pergamon Press, NY., 211 (1982).
7. R.P. Wool and K.M. O'Connor, J. Appl. Phys., 52, 5953 (1981).

8. R.P. Wool and K.M. O'Connor, J. Polym. Sci., Letters, 20, 7 (1982).

9. Y.-H. Kim and R.P. Wool, Macromolecules, 16, 115 (1983).

10 P.-G. deGennes, J. Chem. Phys., 55, 572 (1972).

11. M. Doi and S.F. Edwards, J. Chem. Soc., Faraday Trans. II, 74, 1789 (1978).

12. R.P. Wool, J. Elast, and Plastics, 17, 106 (1985).

13. (a) R.P. Wool, Rubber Chem. Technol., 57(2), 307 (1984).
(b) J.D. Skewis, Rubber Chem. Technol., 39, 217 (1966).
(c) W. G. Forbes and L.A. McLeod, Trans. Inst. Rubber Ind., 30(5), 154 (1958)
(d) G.R. Hamed and C.-H. Shieh, J. Polym. Sci., Polym. Phys. Ed., 21, 1415 (1983).

14. R.P. Wool and A.T. Rockhill, J. Macromol. Sci.-Phys., B20, 85 (1981).

15. J.L. Willett, K.M. O'Connor and R.P. Wool, J. Polym. Sci., Phys. Ed., 24, 2583 (1986).

16. P.J. Flory, Statistical Mechanics of Chain Molecules, Wiley Intersci., New York (1969).

17. W.W. Graessley, Adv. Polym. Sci.,16, 1 (1974).

18. J.D. Ferry, Viscoelastic Properties of Polymers, 3rd Ed., John Wiley and Sons, New York (1980).

19. H. Tadokoro, Structure of Crystalline Polymers, John Wiley and Sons, New York (1979).

20. J. Klein, Macromolecules, 11, 852 (1978).

21. K. Kremer, Macromolecules, 16, 1632 (1983).

22. R.P. Wool, A.C.S. Preprints, 26(2), 139 (1985).

23. S.M. Aharoni. Macromolecules, 16, 1722 (1983).

24. M. Daoud, M.P. Cotton, b. Farnoux, G. Jannick, G. Sarma, H. Benoit, R. Duplessix, C. Picot, and P.-G. deGennes, Macromolecules, 8, 804 (1975).

25. P.G. deGennes, Macromolecules, 9, 587 (1976).

26. G.L. Pitman and I.M. Ward, Polymer, 20, 895 (1979).

27. R.P. Wool, Proc. of IX Intl. Congress of Rheology, Mexico, 3, 573 (1984).

28. C.M. Rimnac, J.A. Manson, R.W. Hertzberg, S.M. Webler and M.D. Skibo. Macromol. Sci.-Phys., B19, 351 (1981).

29. K.M. O'Connor, Ph.D. Thesis, "Crack Healing in Polymers," University of Illinois (1984).

30. A. Y. Lee and R.P. Wool, Macromolecules, 19, 1063 (1986).

31. A.Y. Lee and R.P. Wool, Macromolecules, in press.

Author Index

Volume 1	Homogenization and effective moduli of materials
Volume 2	Oscillation theory, computation, and methods of compensated compactness
Volume 3	Metastability and incompletely posed problems
Volume 4	Dynamical problems in continuum physics
Volume 5	Theory and applications of liquid crystals
Volume 6	Amorphous polymers and non-Newtonian fluids

Tables of contents from other volumes from the program in Continuum Physics and Partial Differential Equations

Homogenization and effective moduli of materials

October 22 - October 26, 1984

J. L. Ericksen
D. Kinderlehrer
R. Kohn
J.-L. Lions

Conference Committee

Theory and applications of liquid crystals

January 21 - 25, 1985

J. L. Ericksen
D. Kinderlehrer

Conference committee

Berry, G.	Rheological and rheo-optical studies with nematogenic solutions of a rodlike polymer: a review of data on poly (phenylene benzobisthiazole)
Brezis, H.	Singularities of liquid crystals and S^2 - valued mappings
Capriz, G. and Giovine, P.	On virtual inertia effects during diffusion of a dispersed medium in a suspension
Choi, H. I.	Degenerate harmonic maps and liquid crystals
Cladis, P.	A review of cholesteric blue phases
Cohen, R., Hardt, R., . Kinderlehrer, D., Lin, S.-Y., and Luskin, M	Minimum energy configurations for liquid crystals: computational results
Di Benedetto, E.	The flow of two immiscible liquids through a porous medium: regularity of the saturation
Doi, M.	Molecular theory for the nonlinear viscoelasticity of polymeric liquid crystals
Hardt, R. and Kinderlehrer, D.	Mathematical questions in liquid crystal theory
Huang, C. C.	The effect of the magnitude of the disordered phase temperature range on the given phase transition in liquid crystals
Leslie, F.	Some topics in equilibrium theory of liquid crystals
Leslie, F.	Theory of flow phenomena in nematic liquid crystals
Maddocks, J.	A model for disclinations in nematic liquid crystals
Miranda, M.	Some remarks about a free boundary type problem
Ryskin, G.	Computer simulation of flow of liquid crystal polymers
Sethna, J.	Theory of the blue phases of chiral nematic liquid crystals
Spruck, J.	On the global structure of solutions to some semilinear elliptic problems

Oscillation theory, computation, and methods of compensated compactness

April 1 - April 4, 1985

C. Dafermos
J. L. Ericksen
D. Kinderlehrer
M. Slemrod

Conference Committee

Chacon, T. and Pironneau, O.	Convection of microstructures by incompressible and slightly compressible flows
DiPerna, R.	Oscillations in solutions to nonlinear differential equations
Forest, M.G. and Lee, J.-L.	Geometry and modulation theory for the periodic nonlinear Schrödinger equation
Harten, A.	On high-order accurate interpolation for non-oscillatory shock capturing schemes
Lax, P.	On the weak convergence of dispersive difference schemes
Majda, A.	Nonlinear geometric optics for hyperbolic systems of conservation laws
McLaughlin, D.	On the construction of a modulating multiphase wavetrain for a perturbed KdV equation
Nunziato, J., Gartling, D. Kipp, M.	Evidence of nonuniqueness and oscillatory solutions in computational fluid mechanics
Osher, S. and Chakravarthy, S.	Very high order accurate T V D schemes
Rascle, M.	Convergence of approximate solutions to some systems of conservation laws: a conjecture on the product of the Riemann invariants
Schonbek, M.	Applications of the theory of compensated compactness
Serre, D.	A general study of the commutation relation given by L. Tartar
Slemrod, M.	Interrelationships among mechanics, numerical analysis, compensated compactness, and oscillation theory
Venakides, S.	The solution of completely integrable systems in the continuum limit of the spectral data
Warming, R. and Beam, R.	Stability of finite-difference approximations for hyperbolic initial value boundary value problems
Yee, H.	Construction of a class of symmetric T V D schemes

Metastability and incompletely posed problems

May 6 - May 10, 1985

S. Antman
J. L. Ericksen
D. Kinderlehrer
I. Müller

Conference Committee

Antman, S. and Malek-Madani, R.	Dissipative mechanisms
Ball, J.	Does rank-one convexity imply quasiconvexity?
Brezis, H.	Metastable harmonic maps
Calderer, M.	Bifurcation of constrained problems in thermoelasticity
Chipot, M. and Luskin,M.	The compressible Reynolds lubrication equation
Ericksen, J.	Twinning of crystals I
Evans, L. C.	Quasiconvexity and partial regularity in the calculus of variations
Goldenfeld, N.	Introduction to pattern selection in dendritic solidification
Gurtin, M.	Some results and conjectures in the gradient theory of phase transitions
James, R.	The stability and metastability of quartz
Kenig, C.	Continuation theorems for Schrodinger operators
Kinderlehrer, D.	Twinning of crystals II
Kitsche, W., Müller, I., and Strehlow, P.	Simulation of pseudoelastic behaviour in a system of rubber balloons
Lions, J.-L.	Asymptotic problems in distributed systems
Liu, T.P.	Stability of nonlinear waves
Maderna, C., Pagani, C., and Salsa, S.	The Nash-Moser technique for an inverse problem in potential theory related to geodesy

Mosco, U.	Variational stability and relaxed Dirichlet problems
Pitteri, M.	A contribution to the description of natural states for elastic crystalline solids
Rogers, R.	Nonlocal problems in electromagnetism
Vazquez, J.	Hyperbolic aspects in the theory of the porous medium equation
Vergara-Caffarelli, G.	Green's formulas for linearized problems with live loads
Wright, T.	Some aspects of adiabatic shear bands

Dynamical problems in continuum physics

June 3 - June 7, 1985

J. Bona
C. Dafermos
J. L. Ericksen
D. Kinderlehrer

Conference Committee

Amick, C. and Kirchgässner, K.	Solitary water-waves in the presence of surface tension
Beals, M.	Presence and absence of weak singularities in nonlinear waves
Beatty, M.	Some dynamical problems in continuum physics
Beirao da Veiga, H.	Existence and asymptotic behavior for strong solutions of the Navier Stokes equations in the whole space
Bell, J.	A confluence of experiment and theory for waves of finite strain in the solid continuum
Boczar-Karakiewicz, B., Bona, J., and Cohen, D.L.	Shallow water waves and sediment transport
Chen, P.	Classical piezoelectricity: is the theory complete?
Glassey, R. and Strauss, W.	On the dynamics of a collisionless plasma
Keller, J.	Acoustoelasticity
McCarthy, M.	One dimensional finite amplitude pulse propagation in electroelastic semiconductors
Morawetz, C.	Weak solutions of transonic flow by compensated compactness
Müller, I.	Extended thermodynamics of ideal gases
Pego, R.	Phase transitions in one dimensional nonlinear viscoelasticity: admissibility and stability
Roytburd, V. and Slemrod, M.	Dynamic phase transitions and compensated compactness
Shatah, J.	Recent advances in nonlinear wave equations
Spagnolo, S.	Some existence, uniqueness, and non-uniqueness results for weakly hyperbolic equations in Gevrey classes

9 780387 965567